D1368036

Fight or Flight

Other Avon Books by
Geoffrey Regan

BLUE ON BLUE: A HISTORY OF FRIENDLY FIRE
SNAFU: GREAT AMERICAN MILITARY DISASTERS

Fight or Flight

GEOFFREY REGAN

AVON BOOKS NEW YORK

FIGHT OR FLIGHT is an original publication of Avon Books. This work has never before appeared in book form.

AVON BOOKS
A division of
The Hearst Corporation
1350 Avenue of the Americas
New York, New York 10019

Published by arrangement with the author
Library of Congress Catalog Card Number: 95-41402
ISBN: 0-380-78019-4

Library of Congress Cataloging in Publication Data:
Regan, Geoffrey.
Fight or flight / Geoffrey Regan.
p. cm.
Includes bibliographical references and index.
1. Psychology, Military. 2. Fear. 3. Military history. I. Title.
U22.3.R44 1996 95-41402
355'.001'9—dc20 CIP

First Avon Books Trade Printing: April 1996

Printed in the U.S.A.

OPM 10 9 8 7 6 5 4 3 2 1

For Annie and Alan Morrison

Contents

INTRODUCTION 1

HEROES AND VILLAINS 11

A SOLDIER'S MORALE 45

FIGHT OR FLIGHT

1 *Fighting Spirit and Fanaticism in Medieval Warfare—The Battle of Nicopolis, 1396 and the Siege of Jerusalem, 1099* 75

2 *The "Bloodybacks" at the Battle of Minden, 1759* 86

3 *"That Astonishing Infantry" at the Battle of Albuera, 1811* 98

4 *"Victory or Death" at the Alamo, 1836* 106

5 *"The Thin Red Line" at the Battles of Balaclava and Inkermann, 1854–55* 114

6 *The 20th Maine at Little Round Top—The Battle of Gettysburg, 1863* 127

7 *"The Legend of Johnny Reb"—The Battle of Missionary Ridge, 1863* 138

8 *"Pals" on the Somme—1 July, 1916* 154

9 *The Battle of Caporetto, 1917—A Monologue in One Act* 173

10 *"With Our Backs to the Wall"—Ludendorff's Offensive, 1918* 179

11 *The "Leathernecks" at Belleau Wood, 1918* 202

12 *"Run, Run, the Bogeyman Is Coming!"—The Fall of Anual, 1921* 209

13 *"A Nation in Arms"—The Fall of France, 1940* 212

14 *The End of "Fortress Singapore," 1942* 222

15 *"Tackling the First Team"—The Battle of Kasserine Pass, 1943* 234

16 *"Tell It to the Marines"—The Invasion of Tarawa, 1943* 245

17 *Into the Dragon's Lair—The Battle of Hürtgen Forest, 1944* 253

BIBLIOGRAPHY 265

INDEX 269

INTRODUCTION

The only thing I'm not certain about is whether I may get the wind up and show it. I'm afraid of being afraid.

—Captain J. E. H. NEVILLE, 1917

Fear is the common bond between fighting men.

—RICHARD HOLMES, 1985

The worst part of battle was wondering how you were going to behave in front of other people.

—RALEIGH TREVELYAN, 1956

The real enemy was Terror, and all this heel-clicking, saluting, bright brass and polish were our charms and incantations for keeping him at bay.

—ALAN HANBURY-SPARROW, 1932

Whenever one surveys the forces of the battlefield, it is to see that fear is general among men, but to observe further that men are commonly loath that their fear will be expressed in specific acts which their comrades will recognize as cowardice. The majority are unwilling to take extraordinary risks and do not aspire to a hero's role, but they are equally unwilling that they should be considered the least worthy among those present.

—S. L. MARSHALL, *Men Against Fire,* 1947

Fear is so fundamental an aspect of warfare that one wonders why in the past its control did not comprise much of a soldier's training. It is a natural emotion and one that has proved vital in the development of human life. Terror, on the other hand, is a destructive extreme. It indicates a breakdown of control, a breach in self-discipline, and no amount of imposed discipline can remedy its effects. Ideally, those who have succumbed to terror should never have been forced to become soldiers. They should have been weeded out at an earlier stage of personnel selection. Such individuals were clearly not suited to a military career and would have been well advised to seek a less stressful profession. Regarded in this way, the problem of combat breakdown virtually solves itself. And if it were simply a case of square pegs in square holes and round pegs in round holes, I would need to write little more than that, and put everything down to selection procedures. But war is not like other professions and never has been. As many unwilling men have fought in history's battles as willing ones. And one day's hero has sometimes become the next day's villain.

Naturally the more disciplined the soldier, the less likely he is to break down under the pressures of war. The discipline of soldiering, in that sense, is no different from the discipline required for other professional exercises. But just as it is necessary to be very careful in selection procedures for most professions, such care and attention must also be paid to the selection of soldiers. Where this has not taken place too many weak and unsuitable men have been sent into—for them—impossible situations, where they have given way to their personal demons and suffered combat breakdown. In Britain and the United States, standing armies have been unpopular, and citizen armies and conscription were only introduced under the pressures of a war situation. As a result it became necessary for democracies like Britain and the United States to impose severe discipline on men who were recruited or conscripted into the army but did not possess the sort of background or experience that might have been taken for granted in the states where conscription had been the norm for many years. Thus men who had led a softer life in "civvy street" had to endure a more severe regime when under arms, in order to toughen them up. In 1918, Field Marshal von Ludendorff had expressed the view that it was the severity of British discipline in their army that had prevented the British army from succumbing to the morale-sapping pressures of the war. The previous year, the Russian army had mutinied and given up the war, the French army had also experienced serious cases of mutiny, while the Italian army had suffered virtual terminal combat collapse after the defeat at Caporetto. It was, in fact, the collapse of the spirit or morale of the German army in 1918 that led to the breakdown of the German war effort. Ironically, Ludendorff was probably right. During the whole war more than 3,080 British soldiers were sentenced to death for cowardice or desertion by courts-martial and some 312 men are known to have been executed by firing squad—by far the highest number of judicial executions for such military offenses of any of the combatants. Moreover, British officers and NCOs were swift to apply instant justice, where a man threatened to cause panic through a loss of nerve. But death cannot be the only answer to the problem of fear in battle. Fear should and must be channeled so that its control becomes the first step in becoming an efficient soldier. This book is concerned with that process.

Under stress everyone suffers the chemical reactions in the body that produce "fight or flight" feelings. These may be quite inappropriate in many walks of life and may even cause "panic attacks" in situations where the sufferer is in no physical danger of any kind. However, for the soldier things are different. For him "fight or flight" often means just what it says. However, just as for many civilians, neither fight nor flight may be appropriate responses for him either. Fighting may be too risky and fleeing is absolutely forbidden. In such cases self-control is the better response. It is really mind over matter; *homo sapiens* triumphing over his brute nature. In this book I have tried to examine how fear can be controlled and how morale or fighting spirit may be maximized. In the part titled "Fight or Flight" I have looked at a series of military encounters which have demonstrated, in a variety of ways, the reactions of armies when confronted with extremely stressful circumstances. In some cases the soldiers triumph by showing outstanding courage; in others they succumb to their terrors and panic sets in. The choice of battles is very much my own. I recognize the dangers of attributing courage or cowardice to national groups and so I merely refer my readers to history, which for every national "triumph" has a corresponding "disaster."

According to Lord Moran, in his famous book, *Anatomy of Courage,* there are four degrees of courage, and four types of men who can be measured by that standard of courage. To Moran there were:

1. Men who did not feel fear.
2. Men who felt fear but did not show it.
3. Men who felt fear and showed it, yet continued to do their job.
4. Men who felt fear, succumbed to terror, showed it, shirked their responsibility, and deserted or fled.

There have always been all four kinds of men in the armies of all warring nations. The problem for commanders has been how to identify those men who would give way and run, so that they might be able to prevent them running or use them in a capacity in which they could do least harm. During the difficult days in Britain in 1917 and 1918, when manpower was short, too many of the wrong sort of men were allowed to reach the battle front. An American case, the tragic one of Eddie Slovik in 1944 (*see* "Heroes and Villains") was just such an example of a deeply inadequate man being forced to join the battle line, when he would have been of some use behind the lines in a support capacity.

Moran suspects that his first case—the fearless man, if he exists—is fearless as a result of having a vacant mind, and that much of the courage that we read of in history is perhaps an indication that the brave man lacks imagination. Personally, I doubt this. In his book *The Western Way of War,* Victor Davis Hanson paints a clear picture of ancient Greek phalangists losing control of their natural bodily functions as the enemy bore down on them. Involuntary urination or defecation of this kind was the clearest evidence of a warrior's state of mind, and it is doubtful if soldiers at any stage of history have been entirely free of this fundamental display of animal fear. Nevertheless, Moran's point is not so easily dismissed. The steadiness of eighteenth century soldiers, who stood upright no more than fifty or so paces from the enemy and exchanged volleys without flinching, is hardly comprehensible to late–twentieth century minds. To anyone who gives a moment's thought to the development of warfare, their behavior was contrary to all our instincts of self-preservation. To the uninitiated in the field of military history, armor makes sense, concealment behind trees makes sense, even trenches make sense—anything makes sense that prevents the enemy having free access to the vital parts of the human anatomy. Standing out in the open, apparently inviting the enemy to shoot you, does not make sense. Were these men, who incidentally wore uncomfortable uniforms and powdered wigs, incredibly brave or just incredibly stupid? Did they feel fear at all, or were they part of the fearless category that Moran postulates? The answer, I suppose, is that their drill was so savage that battle itself became a blessed relief, and a release from the stranglehold that the drillmasters had on their freedom of movement. Their imagination, in Moran's words, played no tricks. They drew no picture of danger, for their own undoing. Certainly the French view—in numerous encounters with English troops—has been that if the English had any apprehension of their situation, they would have run away. Before the battle of Agincourt, in Shakespeare's play *Henry V*, the French nobles claim that the English are in such a desperate situation that if they had any imagination at all they would have fled the field. So through "stupidity" King Henry V's Englishmen win one of their greatest victories.

Sir John Fortescue, great historian of the British army, does not agree with Moran's view that there are "fearless men." In his words: "It is, I believe, a fact, that even the bravest man cannot endure to be under fire for more than a certain number of consecutive days, even if the fire be not very heavy." This seems to me a much more realistic appreciation of the problem.

A study of military history confirms that twentieth century warfare ensures breakdown in battle, or combat exhaustion, at some stage. The pressures of modern warfare are beyond the endurance of ordinary mankind. American research has shown that each moment of combat imposed a strain so great that men broke down in direct relation to the intensity and duration of their exposure, just as the average motor vehicle wears out after a certain number of miles. It appears that the "doughboy," G.I., or indeed the Tommy, wore out eventually, by either developing an acute incapacitating neurosis or else becoming hypersensitive to shellfire. There were no exceptions. The average point at which collapse occurred appears to have been in the region of 200 to 240 aggregate combat days. British research has tended to suggest that 400 combat days was possible if the rotation system was used whereby men were used in the front line once, for a limited period consecutively.

During the First World War it has been suggested that an officer could only be said to be "learning the ropes" during this first three weeks at the front, after which he was at his best for just three or four weeks before his usefulness declined to a point such that unless he was relieved, he would become almost totally useless. Richard Holmes has shown that this kind of combat breakdown is not a twentieth century phenomenon. During the Seven Years' War, he points out, a Prussian officer described how Russian gunners at the battle of Zorndorf in 1758 had so completely lost their fighting spirit that they crouched under their guns and let themselves be massacred. On many other battlefields soldiers have been seen to become completely incapacitated, as if their sense had left them. American psychiatrists have shown that the process by which the breakdown occurs is a general slowing down of mental processes and apathy; as far as the victim was concerned the situation was one of absolute hopelessness. Even the influence and reassurance of understanding officers or NCOs failed to rouse the soldiers from their sense of hopelessness. The soldier had become slow-witted, and memory defects became so extreme that the soldier could not be counted on to relay even the simplest verbal order. He remained almost constantly in or near his slit-trench and during acute action took no part, trembling constantly. Even in the case of good and effective soldiers the insidious process of breakdown took place, and as they lost their friends or received bad news from home they began to feel that their chances of survival were diminishing minute by minute.

How then is the breakdown resisted? In "A Soldier's Morale" I have studied the subject of morale, which some writers believe is synonymous with "fighting spirit." As F.N. Richardson has shown, morale is dependent on both physical and psychological factors. The failure rate in battle can be reduced, and has been reduced successfully by units with high morale. Richardson shows that the health of the soldier is of preeminent importance. In the case of nineteenth century warfare, for example the British campaign in the Crimea, it is remarkable but true that bacteria killed far more of the soldiers than did the bullets of the enemy. This was even true during some campaigns in the Second World War with malaria and other insect-borne diseases having a serious effect upon the fighting capacity of the British Commonwealth and American troops. Numerous examples show that morale was greatly enhanced by the knowledge, on the part of the soldier, that he had a good medical service behind him if he needed it. During the Second World War, the use of antibiotics, blood transfusions, and other surgical techniques, in addition to fast evacuation of injured men from the danger area to a medical station, enhanced morale considerably.

Second only to the existence of good medical services comes the feeling, on the part of the soldier, that he gets the best food available. Napoleon once said that an army marches on its stomach, and this was true for his *Grande Armée.* For the epicurean Frenchman, food had the biggest single effect on his morale. Richardson shows the importance of tea for the British army. Hot, sweet tea was invariably used as a counter to shock, particularly among the wounded. A further factor that contributed to good fighting spirit was and presumably always will be the opportunities provided for rest and sleep.

In addition to these obvious physical considerations, a soldier will only fight well if he does not feel burdened by mental or psychological stresses. Army life itself can be very stressful to conscripts drawn from civilian backgrounds. It is vital, therefore, that every effort should be made to reduce stress and make use of whatever factors are available to stimulate a strong fighting spirit. Religious faith has always been a motivating factor for soldiers and it has been said that there are "no atheists in foxholes." Indeed, many soldiers turn to God in the extremity of battle. Religious fanaticism has frequently been a substitute for good morale in armies of the East, and has sometimes enabled soldiers to overcome apparently impossible odds.

With the decline of religion in Western democracies, morale has been maintained by reference to "the cause," sometimes a pseudophilosophical or a quasi-political one. One thinks of the soldiers of the Confederacy during the American Civil War, with their belief in the cause of the South and the slave system. A moment's rational thought would have shown them that slavery, as a cause, was not worth fighting for and what they were in fact fighting for was a system that violated America's democratic traditions. The fanaticism of Oliver Cromwell's "Ironsides" during the English Civil War was partly made up of religious conviction—that they were fighting against a papist king—and partly made up of a feeling that the cause of parliament was democratic and more in keeping with the lives of ordinary people. Cromwell himself believed his soldiers were motivated by a knowledge that they were fighting for something which was worthy and therefore fought better. The problem, however, for the ordinary soldier was to find out exactly what his "cause" actually was. In 1914 many British civilians volunteered to join the army which was fighting against Germany on a number of curious grounds. Propaganda had made it impossible for the ordinary soldier to understand the complicated diplomatic maneuverings that had brought about the war itself. It was far simpler for them to believe that they were fighting for the future of their own country—a country which was viewed through rose-tinted glasses, of cricket fields and country lanes—and that the English way of life was being threatened by the "Huns." Others believed they were liberating the "poor Belgians" who had been overrun by the Germans. Whatever the truth, most people were susceptible to propaganda and soldiers fought bravely because they felt that it was their duty to fight for "King and Country." In the twentieth century, with a far more sophisticated soldiery, the influence of the "cause" has waned. According to General Wavell, "a man does not flee because he is fighting in an unrighteous cause; he does not attack because his cause is just."

Lord Wavell believed that soldiers fought best when they were part of a good unit. For the ordinary soldier, the good unit is the man's military home and family, and while his real home and family may be very far away, he is able to take a substitute on the battlefield and feel that he is fighting for a unit of which he feels a valued part. The "good" soldier is the soldier who has been well trained and is confident in the leadership of his officers. The officers, in turn, must understand the personalities and abilities of the individual men under their control. Nothing has brought about a breakdown in morale and fighting spirit more quickly than expecting too much of unsuitable troops. Good training and an understanding of what would be expected of him is the greatest assurance a soldier can have.

In conclusion, one must never forget that even the most successful soldiers began their military careers as frightened and uncertain recruits or conscripts. They may have learned to control their fears, helped by the many methods in which morale may be boosted, and they may have flourished through the *esprit de corps* of their regiment or military unit. Nevertheless, history is not the solitary preserve of the most successful, the splendid, and the best. Military history written like that would be a sham. Many misfits pass through the system untouched by the glory that is the reward of the few. In the massed armies of the twentieth century there were many faceless men, hoping against hope that they could survive the ordeal of war and get back home to the lives that they had chosen for themselves. There were thousands of Eddie Sloviks in every army. As British MP Ernest Thurtle said during the debate on capital punishment in the army in 1928, "It is not fair to take a man from a farm or factory, clap a tin hat on his head, and then shoot him if his nerve fails." After all, in the words of Brigadier General Sir John Smyth, "One's natural instinct when shooting starts is to lie in a ditch and stay there until it is all over; and it is only through discipline and training that one can make oneself get out and go forward."

This book is written with understanding of all those who never could get out of the ditch, and stayed there and suffered for it.

Heroes and Villains

The British comedian Spike Milligan described himself in one of his books as being "a hero with coward's legs." Nor is that such a ridiculous description. It may well have applied to more men in history than Spike Milligan can have imagined. President Abraham Lincoln said much the same about Union deserters whose names were presented to him for confirmation of their death sentences. In many cases he refused to confirm the findings of the courts-martial on the grounds that they were "leg cases." Every man has a breaking point. Those who break quickly, succumb to their fears, and run away from battle are called cowards and have always received the strictures of society. Those who have a slower breaking process may, if rested, put off nervous collapse for some time; they may even become heroes. The difference between the two men is merely a matter of time. For the truth is that in combat conditions, every man will break eventually. The intensity of modern warfare—notably in terms of sustained artillery bombardment—makes combat collapse certain, in a way that was never true of the wars of the past. Even as bloody a battle as Waterloo was over in a matter of hours, its intensity stunning but its duration bearable by the common soldier. Few men at Waterloo had time to reflect on their fears; they were too busy trying to preserve their own lives against active and savage foes.

Why do men suffer nervous or psychiatric collapse as a result of combat experience? And why has such collapse become so prevalent in the twentieth century? There are many possible reasons, ranging from the type of man who fights to the type of weapons he uses or has used against him. The term "shell shock" applies to a type of nervous collapse rather than merely to the cases that involve artillery bombardment. Nevertheless, it has often been the sheer intensity of the big guns that has relentlessly pushed men toward collapse. However, an important explanation that should not be overlooked is the way in which modern day soldiers are often "alone" on the battlefield. We will see later how many soldiers gain strength from being part of a unit and resist collapse because they do not want to let themselves or their friends down in public. But what happens if nobody is looking? Or you are on an "empty" battlefield, cut off from your comrades and forced back on your own limited resources? Before the First World War, soldiers usually had the reassurance of being a part of a larger unit, most of which would be visible to them. However, since the advent of modern technology in the twentieth century, the battlefield has become a very lonely place. Most soldiers are under cover and so it is very difficult for them to see more than one or two of their own colleagues. This has the effect of reducing the confidence they may feel in the unit. During Napoleonic battles, infantry stood shoulder to shoulder in their squares, receiving close orders from their NCOs. This required little thinking on their part. The fighting spirit, in that sense, depended on their feeling a sense of unity and strength as part of a larger unit. The modern soldier, on his lonely battlefield, often with no one to see if he goes forward or stays put, must be his own leader.

How quickly, then, will the combat pressures assume dangerous levels and the desire to run or panic become irresistible. One soldier during the American Civil War understood this lonely feeling very well:

> The truth is, when bullets are whacking against tree trunks and solid shot are cracking skulls like egg shells, the consuming passion in the heart is to get out of the way. Between the physical fear of going forward, and the moral fear of turning back, there is a predicament of exceptional awkwardness, from

which a hidden hole in the ground would be a wonderfully welcome outlet.[1]

But what happens when you panic? The German military historian Elmar Dinter defines panic as groups of people fleeing aimlessly, running round in circles, shooting or looting and continuing either until someone intervenes or until complete physical exhaustion is reached. Crucial in this context is the collective abandonment of all self-control. Discipline, whether internally or externally applied, has broken down. From the military historian's point of view, two particular kinds of panic can be observed: one is panic of the group, the other panic of the individual.

As we will see in the next section, there are numerous things that affect adversely the morale and, therefore, the performance of the soldier in battle. Failure to eat regularly, take adequate rest, and maintain general fitness are merely some of the most obvious. These factors will contribute to a state of stress; in this state of stress, reaction to it is involuntary, and the chemistry of the body takes over, notably the adrenal glands, the hypothalamus, and the pituitary gland, which release stress hormones—adrenaline and noradrenaline, and other adrenaline hormones, in particular, cortizol, thereby stimulating the blood pressure and circulation. This is the "fight or flight" syndrome, designed to improve the individual's chances of survival by either maximizing his fighting potential against an imagined or real enemy or by concentrating his abilities to escape from danger by running away. The intention is to make survival more likely. Stress is a form of stimulation and can be positive as well as negative. However, although these reactions to stress are perfectly natural, they are unacceptable in the modern military situation—fight or flight is no longer a realistic approach to the modern world. The fight or flight syndrome is activated by a feeling of being threatened and the stress mechanism is activated directly once one feels threatened. However, the fight or flight syndrome

[1] *Battles and Leaders of the Civil War, Volume 2,* p. 62.

originally developed as a natural reaction to circumstances that existed in our early formative stages and many of them are no longer relevant to the modern world. As Dinter has pointed out, we are not immediately threatened by the sight of somebody with a rifle three hundred meters away; we are a great deal more threatened by somebody with a large club only twenty or fifty meters away. The result is that our reasoning powers interpret a situation which our natural body ignores. Thus, although the danger of the person with the rifle is much greater than the man with the club, our reaction will be very much slower to the rifle than it will be to the club. Once we have suffered a hit from a rifle at three hundred meters, if we survive, we will at least have learned to take it far more seriously. As Dinter has shown, stress enhances the capacity for physical combat against an individual perceived but reduces the human capacity to make reasoned judgments about potential threats. Much stress reaction is entirely futile; running away from a perceived threat may well be pointless and actually far more dangerous than standing still or taking cover, thus flight as a primitive response is unsuitable in a more sophisticated military situation. In fact, as with the trench warfare of the First World War, movement either forward or backward was extremely inadvisable and it was necessary to suppress the effects of the "fight or flight" stress by staying exactly where you were in spite of the extreme chemical stimulation within the body. The suppression of this stress had a harmful effect on the health of the soldier and contributed to combat exhaustion and breakdown.

If all men in battle are subject to chemical stimulation within the body over which they have no control, why do some men flee the field while others fight, more or less appropriately? In *Fight or Flight* we will see many examples of military engagements that brought out the best or the worst in military units. But first we will look at how a number of individuals reacted to combat situations.

THOSE WHO FOUGHT

In *The Anatomy of Courage,* Lord Moran has described courage as "a moral quality," not "a chance gift of nature like an aptitude for games. It is a cold choice between two alternatives, the fixed resolve not to quit; an act of renunciation made not once but many times by the power of the will. Courage is will power."

History had recorded and celebrated far more cases where men displayed courage than those where they displayed cowardice, and this is probably how it should be. Perhaps we should "accentuate the positive," but the historian must always resist the tendency to "eliminate the negative." In this section I intend to investigate cases of men who—presumably—experienced the same physical sensations as any humans do when confronted by a fearful situation. However, they responded in entirely different ways. Some, like the Australian Albert Jacka and the American Alvin York, responded with courage and selflessness and are rightly remembered as heroes. Others, like Eddie Slovik or Edwin Dyett, succumbed to their fears and acted in a way that society deemed reprehensible. Jacka and York were rewarded with medals; Slovik and Dyett's actions cost them their lives.

The Australian Imperial Force—which saw service in Gallipoli and on the Western Front during the First World War—was an extraordinary organization. Comprised entirely of volunteers, it possessed a higher proportion of independent-minded men than perhaps any similar force in history, even including the Confederate Army of Robert E. Lee. Fighting generally under British officers and within the administrative umbrella of the British Army, it rejected much of the traditional discipline of the British and substituted instead a relaxed—some even feared "carefree"—approach to matters military. However, when there was fighting to be done, the Australian soldier took second place to nobody. The Germans always knew the British were planning an attack when Australian units were moved into the line. Difficult and insubordinate as many of the Australians could be on occasions, they produced some of the most outstanding examples of group courage on record. But they also produced many individual heroes, of whom Albert Jacka was one of the most famous.

When Albert Jacka reached Gallipoli in 1915 he was just a private soldier, but his rise was meteoric. By August he was a lance corporal, a full corporal the next day and sergeant by September. In November he was a company sergeant major. By April 1916, Jacka was an officer. This rise from the ranks might have unnerved some men, but Jacka had a natural authority that made others accept him at whatever level he reached.

Jacka won the Victoria Cross on 19 May, 1915, during his service against the Turks at Gallipoli. Jacka's unit—the 14th Battalion—throughout the AIF known as "Jacka's Mob," was holding a position at Courtney's Point. When the Turks overran his position Jacka single-handedly recaptured the trench from a dozen Turks, shooting five of them and bayoneting two others. While the rest of the Turks fled, Jacka lit a cigarette and reported, "I managed to get the beggars, sir" to his company commander. Jacka was the first Australian to win the Victoria Cross in the First World War and he became the idol of the AIF, a legend in his own lifetime. Sergeant E.J. Rule described his impression of Jacka:

> To me, Jacka looked the part; he had a medium-sized body, a natty figure, and a determined face with a crooked nose. His feat of polishing off six Turks single-handed certainly took some beating. At that time one characteristic above all endeared him to all the underdogs: instead of criming men [putting them on a charge] and bringing them before the officers, his method was: "I won't crime you, I'll give you a punch on the bloody nose." His leadership in his last battle was as audacious and capable as in his first . . . Not we only, but the brigade and the whole AIF came to look upon him as a rock of strength that never failed. We of the 14th Battalion never ceased to be thrilled when we heard

> ourselves referred to in the *estaminets* or by passing units on the march as "some of Jacka's mob."[2]

"Jacka's Mob," incidentally, was renowned for its "possession of the least polish but the most numerous war scars."

Albert Jacka was not one to rest on his laurels. He and his "mob" took part in the great battle of the Somme in 1916. Unsurprisingly, Jacka's heroism on this occasion, actually turned the tide of battle in a most extraordinary way. During the fighting around Pozières on 8 August, 1916, a German counterattack had overrun the Australian positions, capturing a company of the 48th Battalion and putting two battalions to flight. One of these battalions was the 14th—"Jacka's Mob"—and Second Lieutenant Albert Jacka, V.C., was in command of the 5th Platoon of the 14th. Jacka's dugout, in fact, had been bombed by the Germans, and several of his colleagues wounded. Jacka tried to gather together all his fit men to fight their way back toward the British lines, but as he emerged from the dugout, with just seven fit men, he found himself surrounded by Germans, who were marching away the Australian prisoners from the 48th Battalion. Jacka and his men immediately charged into the Germans but each of them was shot down by German fire. Although wounded, Jacka was back on his feet in a moment and began to rally Australians from other nearby positions. Some of the men of the 48th grabbed weapons from their guards and joined the melee. Jacka was in the thick of everything and was wounded seven times by bullet and bayonet. It is estimated that he killed twenty Germans himself, while more and more Australians came running to join in. Suddenly, a fresh platoon of Australian soldiers reached the scene and the Germans threw down their arms and surrendered. Jacka's counterattack "against an overwhelming and triumphant enemy" was as brilliant as it had been unexpected. All the prisoners from the 48th Battalion were freed, all the German guards were

---

[2]P. Charlton, *Pozières 1916*, p. 214.

killed or captured, and a further forty-two German soldiers were taken prisoner into the bargain. Incredibly, all the Australian casualties were collected up except Jacka himself. As he later told Sergeant Rule, "A stretcher bearer came, took off my tunic, and fixed me. I asked him to go and bring a stretcher. He went away and I never saw him again. I lay there for a long time, and then began to think of the wounded that were never found. I made up my mind to try to get back by myself." In fact, after all his exertions, Jacka crawled most of the way back to the British lines before he was eventually found. The men who found him did not think he would survive his wounds. One of them told Rule, "The bravest man in the Aussie Army is on that stretcher . . . It's Bert Jacka and I wouldn't give a Gyppo piastre for him; he is knocked about dreadfully." But true to form Jacka recovered and returned to his beloved 14th Battalion. He should have had another V.C. but instead received a Military Cross.

The Official Australian Historian, CEW Bean, described Jacka's counterattack as "the most dramatic and effective act of individual audacity in the history of the AIF." In simple terms, Jacka had turned the tide of battle, snatching victory from defeat. Yet, as so often it seemed with the unconventional Australians, Jacka's courage was merely taken for granted. The War Diary for General Gough's Reserve Army for that day simply recorded: "About 5 A.M. the enemy attacked our lines N of the Windmill on the E. of Pozières: they effected a lodgement, but were driven out. An attack N. of Pozières about the same time by the enemy received the same treatment. Some prisoners were taken in each case." What Captain Jacka, V.C., would have thought of this we do not know. He probably just laughed into his amber nectar [beer].

Even now Jacka was not finished. At the battle of Messines in 1917, he led his company to capture German machine gun positions and personally took a German field gun. At the battle of Polygon Wood, "Jacka's Mob" overran German positions with Captain Jacka at their head. In the opinion of most commentators, Jacka should have won the Victoria Cross three times! He may well have been the bravest man in the entire war. But Jacka's courage was faintly un-British. It was all too flamboyant and demonstrative. It smacked of "collecting medals." So Jacka never received from the British the amount of recognition he deserved. But the British were wrong. Jacka was a scrapper; he seemed to enjoy the "rough and tumble" of warfare and had a nerveless quality that made him pursue his impulses. He valued his life as much as any man—his crawl back to safety shows that—but he was not the sort of man who would allow self-doubt to prevent him from helping out his mates or his "side." Jacka was a patriot but not an unquestioning one. He was unconventional, following his own rules of combat, and not those devised by a desk-bound officer who never fought twelve Turks at the same time and won!

Heroism for German soldiers was so tied up with the concept of duty, that it is sometimes almost impossible to separate one from the other. Yet there were as many occasions when individual German soldiers exceeded either their duty or what could reasonably have been expected of them as there were when Britons, Australians, or Americans showed outstanding bravery. Two examples from the Second World War illustrate moments when individual German soldiers turned the tide of the fighting or even won battles single-handedly.

Hugo Primozic had been a locksmith before joining the *Wehrmacht* in 1934, and it was this capacity for precise work combined with his remarkable unflappability that psychologists picked out in him during his early days in the horse artillery. It was clear that Primozic had outstanding leadership qualities and the capacity to inspire confidence in other men. These were important military virtues and it was not long before he was working as an instructor in armored warfare. However, instead of tanks, Primozic made assault guns his special expertise. These massive cannons, used in support of the tanks and often required to take on enemy tanks at close range, required men of courage and great skill. Primozic soon became the best. In five months on the Russian front in 1942, he won all available German awards for bravery, from the Iron Cross Second Class to the Iron Cross with Oak Leaves, the first German NCO ever to win such an award. In gladiatorial combat he faced Russian tanks in his great armored gun every day. For three consecutive days in September 1942, he destroyed five Russian tanks. On the fourth day he was given the task of guarding the flank of one of the German divisions. He was so busy that he used up all his ammunition. Some distance away he noted another assault gun that had been disabled in the fighting and which contained four soldiers, now helpless to defend themselves. With Russian troops just a hundred yards away, Primozic roared forward under heavy fire. Stopping alongside the wrecked gun he leaped out, found two steel hawsers, attached them to the wreck, and then hauled it back to German lines.

Primozic returned to the front line in company with a second assault gun. These two alone were expected to hold back masses of Russian armor advancing in the vicinity of Rshev. Under heavy bombardment from the Russian artillery, the second assault gun was hit and destroyed. Left alone, Primozic began a "turkey shoot" of advancing Russian tanks. At a range of no more than a hundred yards he took on the whole of the elite Stalin Tank Corps, consisting of fifty tanks. In the first furious phase Primozic destroyed seven of the Russian tanks, then, as the heavily armored vehicles re-formed and attacked again, he disabled a further seventeen inside an hour. Before dark he had hit seven more, making a day's total of thirty-one tanks. By the end of the following day his incredible haul was thirty-nine Russian tanks destroyed. The entire Russian offensive at Rshev was blunted and the German High Command credited Primozic with having won the battle of Rshev single-handedly. If this may seem like an exaggeration, it should be borne in mind that each of his encounters with a tank required tactical skill. Few of them ended with a single shot and sometimes he did not gain his victory without an epic struggle. His other crewmen, notably his driver, must deserve some praise, but ultimately it was Primozic's deadly firing that won the day. His victory was not achieved without some desperate moments. Once the caterpillar track was shot away from his vehicle. While this was being repaired the gun had to remain stationary, apparently a sitting duck for the Russians. But Primozic held them off until the gun was mobile again.

Primozic was commissioned for his exploits and was turned into something of a propaganda tool. He was interviewed on German radio, but he was a man of action rather than words. He could not inspire the German masses with lurid descriptions of his victories. When asked to define courage, he was baffled. According to him the vital thing had been to remain calm. He survived the war and was the most decorated of all Germany's fighting men. His secret had been his perfectionism. So single-minded was he in pursuit of his profession that he could allow no time for fear. Fear implied the possibility of failure. To succeed he could entertain no doubts.

The name of Captain Michael Wittman, of the 501st SS Heavy Tank battalion who "won" the battle of Villers-Bocage and blunted Field Marshal Montgomery's push toward Caen on 13 June, 1944 should be far better known. Although the battle of Villers-Bocage represents a low point for the British Army in the Normandy campaign, as well as a particularly sad period for the 7th Armored Division—part of Montgomery's immortal "Desert Rats"—this should not be allowed to detract from Wittman's extraordinary achievement. The division's own historian agrees that "the normally very high morale of the Division fell temporarily to a very low ebb. . . . A kind of claustrophobia affected the troops." Whatever it was that affected the troops, they lost their sharpness and their almost-lackadaisical attitude was noted and exploited by the Germans.

Field Marshal Montgomery's plan to take Caen had originally allowed for an airborne drop by 1st Airborne Division behind the town as soon as 7th Armored and 51st Highlanders had pushed up from the coast. However, the complete failure of 7th Armored to make that push put an end to such planning. Major General Erskine, commanding the 7th Armored Division, had initially been excessively optimistic, reporting limited German resistance and the loss of just four tanks. Yet the division was moving far too slowly and was failing to give armored support to the infantry. It was on 11

June that XXX Corps commander, General Bucknall, detected a large gap in the German defenses between Caumont and Villers-Bocage. It was just asking for the British armor to rip through. The next day 7th Armored was ordered to swing westward to launch an attack on the enemy's left and exploit the gap. The forward tanks of 7th Armored moved to within six miles of the hilltop town of Villers-Bocage before stopping for the night.

The next morning—13 June—the leading tanks of 7th Armored entered Villers-Bocage, greeted with transports of delight by the French citizens. So unexpected was the British arrival that German billeting officers were still visiting houses to find accommodation for German soldiers in the town. Suddenly there was an air of festivity and some of the British crews dismounted, as if confident there were no Germans in the vicinity. They began chatting to the local people and accepting flowers and cakes. Montgomery, delighted by news of 7th Armored's progress, signaled their success to England. But everyone was being lulled into a false sense of security. Unknown to them a single Tiger tank commanded by the leading tank "ace" of the Second World War, Captain Michael Wittman, was approaching. Wittman stopped his tank and surveyed the British columns ahead of him. It seemed like peacetime maneuvers, with British tank crews brewing up and enjoying the pleasant sunny weather. He was amazed at how complacent they all were. "They're acting as if they've won the war already," said Wittman's gunner. The great man looked down and replied, "We're going to prove them wrong."

Suddenly Wittman's Tiger roared toward the stationary British Cromwell tanks. One after another Wittman pumped shells into each of the British tanks, leaving them flaming wrecks. Bursting through the one remaining Cromwell, Wittman emerged in the main street of Villers-Bocage like one of the horsemen of the Apocalypse. He next met three more tanks of the County of London Yeomanry and destroyed each of them, then missed another Cromwell, which disappeared into a garden somewhere. A British Sherman hit Wittman's Tiger several times without penetrating its armor. Wittman replied by blowing apart a building behind which the Sherman was sheltering, deluging the tank with its debris. Wittman then destroyed one more Cromwell and reversed away to reload and refuel.

Meanwhile, the rest of "A" squadron, commanded by Lieutenant Colonel Lord Cranley, was being attacked by four other Tigers from Wittman's command. Within a short space of time Cranley's tanks had been destroyed and himself taken prisoner. Wittman, having crushed the vanguard of 7th Armored, now joined other Tiger tanks from the 2nd Panzer Division in their attacks on the British forces around Villers-Bocage. However, Wittman's luck had run out and the British ambushed him, destroying his Tiger, though he made his escape unhurt. But with more German troops moving into the area, 7th Armored were forced to pull back what remained of their spearhead to the town of Tracy-Bocage, two miles away to the west. In one morning—and mainly at the hands of Wittman himself—they had lost twenty-five tanks and twenty-eight armored vehicles. It had been a shocking defeat. German general Fritz Kraemer, with due understatement, acknowledged Wittman's unprecedented achievement: "The idea of placing in readiness the five Tiger tanks which had been left intact during the enemy air attacks had produced good results." An RAF bombing attack had to be called in to flatten Villers-Bocage. On the following day 7th Armored's progress was again held up by the devastating fire of the handheld *Panzerfaust* antitank gun. Without adequate infantry support—inexplicably Bucknall had failed to ask Second Army for reinforcements—the British tanks were practically helpless against this weapon.

The British had failed to take the opportunity they had been offered and had allowed the Germans to close the gap in their lines that 7th Armored had been aiming to exploit. British tank tactics had been amateurish throughout, with forward units traveling far ahead of any infantry support. The result was that whole columns of tanks could be held up by small groups of German infantry with antitank guns. The tank's high explosive shells were quite inappropriate for use against infantry either well dug in or highly mobile.

During the fighting inside Villers-Bocage, 7th Armored had let Wittman run rings round them, destroying tanks at will. There were now serious questions being asked about the division's fighting spirit. There was a suspicion that many of the tank crews felt that they had done their fighting in the desert against Rommel and that it was somebody else's turn now. Fundamentally such attitudes only flourish under slack and uninspiring leadership. Risk avoidance was the name of the game and battle shyness was strongly built into these veteran units. Part of the trouble was the fact that tank crews had little confidence in their equipment. They felt that the Cromwell and Sherman tanks that they were using were in no way a match for the German Tigers. Even if this was true, there were ways of dealing with Tiger tanks and the feeble display of the tank crews facing Wittman showed just how much influence a spirited individual could have. It was a question of inspired leadership: the Germans had it and the British did not. As far as Montgomery was concerned the fault rested with the divisional and corps commanders, Erskine and Bucknall, and both men were subsequently sacked. As General Dempsey later wrote, "7th Armored Division was living on its reputation and the whole handling of the battle was disgraceful." Wittman had punished the British for their complacency. No war is over until the last bullet is fired. As far as Wittman and men like him were concerned, that last bullet would be fired by Germany. It is interesting to note that before Wittman died, killed near Falaise in early August 1944, he was credited with a personal score of 138 tanks and self-propelled guns as well as 132 antitank guns.

On 8 October, 1918, the Americans were advancing through the Argonne towards Mézières and Sedan. They were being held up by machine gun positions and so they sent out a patrol of twenty men under a sergeant. The patrol eventually stumbled upon a dell that contained seventy-five German troops. At this stage of the war, German resistance was not very strong and these men took the opportunity to surrender to the Americans. However, fate intervened. The machine gunners, whom the Americans had been searching for, began opening fire on the patrol, killing or wounding twelve of the Americans. As it happened, one of these was Corporal Alvin York. York was a backwoodsman from Fentress County, Tennessee, with a reputation as a sharpshooter. Originally York had been a conscientious objector but had been persuaded by his pastor that it was his duty to serve his country. After this York became a clinical killer. From a kneeling position he proceeded to shoot the machine gunners, one after the other. While he was thus engaged a party of Germans led by a young lieutenant charged at York. As they ran toward him, York picked them off one by one until he ran out of ammunition. His experience in turkey-shooting contests in Tennessee stood him in good stead. He drew his .45 automatic and continued killing his assailants until the Germans threw down their arms and surrendered. York stopped firing and collected up his prisoners—of whom there were as many as ninety—marching them back toward the American lines, where he handed them over to a lieutenant. "How many are there, Corporal?" asked the lieutenant. "Jesus, lieutenant, I ain't had time to count them yet." York memorably replied. In fact, by the time he reached his own lines, York had 132 prisoners, driving them all himself with a revolver in the back of the only German officer.

In 1818 an English clergyman decided to confer a small annuity on a British soldier who had shown great courage in the recent battle of Waterloo. The duke of Wellington was asked to choose a worthy recipient for the award and handed on the difficult task to Sir John Byng. Byng decided to make his choice from the 2nd Brigade of Guards, which had distinguished itself in the defense of Hougoumont. There were many gallant candidates, but the election fell on Sergeant James Graham, of the light company of the Coldstream Guards.

The defense of the Château of Hougoumont was one of the decisive moments of the entire battle. Had the French been able to take it, they would have been able to turn Wellington's position on the ridge. Napoleon, however, had underestimated the fighting quality of the British Guards, who defended the Château. Despite being heavily outnumbered and with the buildings burning all around them, they clung on grimly throughout a whole day of the fiercest fighting. At one point as Elizabeth Longford tells us, the French broke open the doors and got inside:

> A gigantic [French] subaltern named Legros and reinforced with the nickname of *L'Enfonceur,* the Smasher, stove in a panel of the great north door and followed by a handful of wildly cheering men, dashed into the courtyard. Pandemonium broke out . . .[3]

In a day of such heroics that individuals could hardly be singled out, Sergeant James Graham was a giant among men. Nearly all the Frenchmen who had forced their way in were killed on the spot; and, as the few survivors ran back, five of the Guards, Colonel Macdonnell, Captain Wyndham, Ensign Gooch, Ensign Hervey and Sergeant Graham, by sheer strength, closed the gate again, in spite of the efforts of the French from without, and effectually barricaded it against further assaults. Over and through the loopholed wall of the courtyard, the English garrison now kept up a deadly fire of musketry, which was fiercely answered by the French, who swarmed round the curtilage like ravening wolves. Shells, too, from their batteries, were falling fast into the besieged place, one of which set part of the mansion and some of the outbuildings on fire. Graham, who was at this time standing near Colonel Macdonnell at the wall, and who had shown the most perfect steadiness and courage, now asked permission of his commanding officer to retire for a moment. Macdonnell replied, "By all means, Graham; but I wonder you should ask leave now." Graham answered, "I would not, sir, only my brother is wounded and he is in that outbuilding there, which has just caught fire." Laying down his musket, Graham ran to the blazing spot, lifted up his brother, and laid him in a ditch. Then he was back at his post, and was plying his musket against the French again before his absence was noticed, except by his colonel. Graham's courage was that

[3] Max Hastings, *The Oxford Book of Military Anecdotes,* p. 235.

of a loyal, dutiful soldier, inspired by the occasion and by his feeling of *esprit de corps.* It is doubtful if even a man like this was without fear. Yet for him it was of no greater significance than any other minor fact of his life as a soldier, like the weather or the quality of army food. He had too much self-respect to show that he was ever afraid.

THOSE WHO FLED

How do potential cowards and deserters find themselves in situations which enable them to give way to their weaknesses? The following two cases—both ending in execution by firing squad—suggest that many such cowards were victims of a selection procedure that put quantity above quality and which willingly sent "hopeless cases" to places where failure was less "possible" than "certain."

The cases of Edwin Dyett and Eddie Slovik

The only American soldier to be executed for cowardice in either world war was Private Eddie Slovik. In all probability he will be the last American soldier ever to pay the supreme price for military inadequacy. Military law decreed that Eddie Slovik had failed as a soldier and so must die. Failure in no other profession carries with it the death penalty. If it did, nobody would dare pursue it except those with an absolute certainty of their ability to cope with its demands. Yet, millions of men of all nations were called upon to be soldiers, even though, for countless personal reasons, they might have been quite unsuited to military life. And so it was with Eddie Slovik. As a petty criminal Slovik had been classified as 4-F by the recruitment board, little above Woody Allen's famous quip of his being a "hostage in wartime." In Slovik's case the classification was fairly accurate. Slovik was not equipped mentally to be a frontline soldier. With what justification did those eventually scraping the barrel in 1944 decide that Eddie Slovik had improved sufficiently to be classified 1-A, (able to be a rifleman and engage the enemies of America in hand-to-hand combat)? Was their decision fair, either to Slovik himself or to the combat unit to which he would be assigned? Was not the true cowardice that on the part of those whose job was to keep the replacement program going, filling gaps in the frontline soldiers with anyone with two hands to hold a rifle and two feet to run away with when the fighting got too hot?

Eddie Slovik was "a loser." He was one of life's victims. Of the forty thousand Americans who were missing from the army in Europe in 1944, just forty-nine were sentenced to death for desertion. Forty-eight of these had their sentences commuted to life imprisonment. A year later they were released. Only Eddie Slovik was executed. Those were long odds. A man could think himself very unlucky—or just fated!

By the time he reached eighteen, Eddie Slovik had a long prison record for minor offenses. As a boy he would steal candy or bubble gum from stores, and once some bread and a cake from his employer. If ever a group of young tearaways broke in anywhere, Eddie got caught, the others got away. He was just a misguided teenager, an educational underachiever. In 1937 he was sent to prison for between six months and ten years for embezzlement. Eddie would not have known the meaning of the word, and he certainly could not spell it. In fact, he had once pocketed the change from a customer when he was working in Cunningham Drug Stores, Detroit; that was his "embezzlement." The "small time" was invented for Eddie Slovik until his one big day—his last. Those who knew Eddie well, and they were mostly his prison guards, understood his limitations. Questioned after the war they were all "wise after the event." As Harry Dimmick, Eddie's reformatory supervisor, said:

> But I could have told the General this. I could have said, General, you can take Eddie Slovik and put a suit on him and teach him how to march in step. You can hand him a

> gun and try to teach him how to shoot it. You can march him up to the battle line and make him dig a hole. But I can tell you *exactly* what he's going to do. When it gets dark, and the first shell bursts over his head, if *somebody stronger than he is is not there to hold him by the hand,* he's sure going to do one of two things. He's either going to freeze right there in the bottom of that hole, or else he's going to jump out and run like a scared rabbit. And there's nothing you can tell him—and not much you can do to prevent it . . . he was a poor risk to stand fast in the dark with guns roaring around him. We knew . . . and the army knew it; that's why they had him in 4-F.[4]

It seems that everyone knew that Eddie Slovik would never make a soldier, yet nobody did anything about it. As a result, about a year after Eddie's wedding to Antoinette Wisniewski on 7 November, 1942—a happy year for Eddie, the best in his life—a letter arrived to spoil everything. Eddie's military classification had suddenly changed, from 4-F to 1-A. Eddie was stunned. His new wife was pregnant and he was just beginning to make a better life for himself when this thunderbolt struck. Why had Eddie undergone such a transformation, from being the least desirable type of recruit to being the most? Was it an error or was the recruitment board getting desperate? The truth was that the board was scraping the bottom of the barrel but could not admit it.

Eddie was taken off for army training. Antoniette had a miscarriage. Eddie hated being away from his wife and wrote her several letters every day. Army life did not suit him. As he wrote, "I wrote a letter this morning and the sergeant gave me hell because I wasn't supposed to write in the morning . . . The way I feel I don't care if they kick me out of this damn army. I'm failing in all my lessons. I'm not thinking of the lessons, just about my wife. You can't blame me, mommy, can you?" In March 1944 he wrote to tell her, "Mommy, I just got off the rifle range and my score was bad but I don't care. When I got down

---

[4]W. B. Huie, *The Execution of Eddie Slovik,* p. 32.

there to shoot I was scared. I tried to tell them that I was nervous but they made me shoot anyway." Eventually, Eddie completed his training. He was as ready for combat as he was ever going to be.

Eddie Slovik sailed from New York on August 7, 1944, aboard the liner *Aquitania,* arriving in Scotland a week later and traveling the length of Britain to undergo the briefest training in combat conditions near the city of Plymouth, in Devon. Then, singularly ill equipped as he was for facing the veteran *Wehrmacht* troops in France, he crossed the English Channel and came ashore on Omaha Beach in Normandy. Neither flag-waving French women nor hard-fighting Panzer troops met him as they had the American assault troops on June 6—D Day—for he was too late for that particular date with destiny. But Eddie had his own D Day awaiting him, less than six months ahead. With eleven other replacements, Eddie had been assigned to "G" Company of the 109th Infantry, part of the 28th Division. He was due to join his company at Elbeuf, but events transpired to thwart his efforts and set in motion a chain of events that was to win him a notoriety denied any other American soldier in the previous eighty years.

Eddie and his companions were taken by truck toward Elbeuf. It was a startling introduction to the realities of modern warfare for the rookies. German troops in the area had been trapped in what became known as the "Falaise Pocket" and massacred by British rocket-firing *Typhoons.* That part of Normandy was like a vast slaughterhouse. It was probably at this moment that Eddie's nerves, never very strong, began to give way. Mile after mile of burning tanks and charred corpses of men and horses lined the road toward Elbeuf. Eddie's friend, John Tankey, recorded his own thoughts of that nightmare journey inland and their arrival at Elbeuf:

> A lot of shells were going over, and we were fired on two or three times. We didn't have an officer, just this one non-com. Close to midnight we got to this open lot and the non-com said dig in. We did. I dug a good deep hole and so did Eddie. A lot of shells were going over. We dug in

> good and deep. As for Eddie, I guess he was scared—I sure as hell was—he was just human. Like I say, we dug these holes and we would yell back and forth to each other.[5]

At that moment tanks were sighted coming toward them and they were not American. A voice began yelling that they were "Krauts." Eddie was the first to recognize that they were Canadian. Eddie and Tankey were relieved, but when they looked around for the rest of their small party they found that they had gone. Disoriented by their experiences, the two men took up the Canadians' offer of joining up with them until they could find their own company. Slovik and Tankey were obviously "missing" but as yet nobody would have regarded them as having deserted. But the shell fire in Elbeuf had brought about an immediate change in Eddie Slovik as Harry Dimmick suggested it would. He had had more than enough of firing and danger for one war: he wanted out.

Ironically, Eddie and Tankey had joined up with part of the Canadian Provost Corps, just eighteen soldiers under the command of a sergeant. The Canadians found the two Yanks "damn good guys," willing to work hard and take on the most unpopular duties. When the Canadians headed back toward the coast, Eddie and Tankey went with them. Unfortunately, the 109th Infantry—the unit Eddie had never reached—was heading in the other direction, toward Paris. For forty-five days the 109th just had to make do without two of its replacements. Aware that they were at least technically in the wrong for not joining up with "G" company, Tankey wrote a letter on behalf of both himself and Eddie Slovik explaining how they had come to get lost. But why did they not follow their letter in person? The answer, it would appear, is that they were enjoying themselves too much with the Canadians. Eddie, for example, displayed previously undiscovered skills in cooking potato pancakes as an alternative to the Canadians' dire diet of "bully"—or boiled—beef. Incredibly, Tankey actually re-

[5]W. B. Huie, *The Execution of Eddie Slovik,* pp. 107–8.

ceived letters from his wife in the United States which reached him via one of the Canadian noncoms. Eddie even learned to ride a motorcycle courtesy of the Canadians. But if this was the kind of war that appealed to Eddie Slovik, it was not why he had been sent to France. While he gave up carrying ammunition, he collected stationery instead and continued bombarding his wife with letters. When not writing home he was at everyone's beck and call. The Canadians liked this amiable Yank, who enjoyed doing favors for everyone and seemed as unwarlike as any soldier they had ever met. He and Tankey even took a German pilot prisoner and infuriated local Belgians by treating him to cigarettes and free drinks.

While Eddie and Tankey took part in their own war, the 109th had endured "triumph and disaster." Had Eddie Slovik reached "G" Company, he would have enjoyed the honor of marching with the rest of 28th Division through the newly liberated capital of France. He would also have taken part in the horrific fighting in the Hürtgen Forest, which cost the division so many casualties. Instead he trailed around with the Canadians, behind the front lines, doing useful things for everyone but not fighting for his country, as he had been trained to do. Yet a part of Eddie Slovik was telling him that what he was doing just was not right and that he should really try to join up with the 109th. Eventually, after an absence of some forty-five days, Slovik and Tankey reached the regimental headquarters of the 109th at Rocherath, in Belgium. Incredibly, the authorities there were willing to swallow their story and seemed unconcerned as to why they were so late. What mattered more was that they had arrived now and were available for immediate action. Eddie and his friend were ordered to join up with Captain Grotte, commanding "G" Company, but from this moment onward Eddie Slovik and his buddy's fortunes diverge. Tankey was prepared to fight for his country; Eddie Slovik was not. Tankey had heard it all before from Eddie but had not taken it seriously. From those first moments in Elbeuf, Eddie had told him that he was not prepared to fire his rifle and that he would run away at the first opportunity. What Tankey had not realized was that the arrival of the Canadians had forestalled Eddie's desertion right there and then. Tankey fought in the Hürtgen Forest and was wounded on 5 November. He assumed that Eddie must be fighting somewhere else. He was quite wrong. No sooner had Eddie met Captain Grotte than he had come out into the open about his true feelings. He told the officer that he was unwilling to fight and would run away if he was put into the front line. Captain Grotte concluded that Eddie was merely panicking and would soon settle down. But Slovik had already made up his mind what he wanted and no amount of persuasion by the captain was going to make him change it. Grotte allocated Eddie to 4th Platoon and warned him not to attempt to leave the company compound. Eddie asked him if he could be tried for being absent without leave. Grotte ordered Eddie to return to his platoon area and stay there. However, after just an hour or so, Eddie returned to see Grotte and asked him, "If I leave now, will it be desertion?" Grotte told him it would and Eddie apparently put down his gun and left his office. Grotte called out to a soldier who happened to be nearby, "Soldier, you better stop your buddy. He is getting himself into serious trouble." The soldier ran after Slovik and pulled him back by the shoulder, but by this stage Slovik would not be thwarted. "Johnny," he said, "I know what I'm doing." The soldier let him go, assuming that he would return. In fact, he was never to see Eddie Slovik again.

There is no clear evidence as to where Slovik stayed that night, but the next morning he surrendered himself to the Military Government Detachment of the 112th Infantry in Rocherath, melodramatically handing a written confession to a cook. The cook took the paper to Lieutenant Griffin of the 112th and Griffin telephoned the 109th and asked them to send someone over to collect the deserter. In this way Eddie returned to his unit, where he was placed under arrest. One of his officers warned him that the "confession" was a dangerous document and suggested that he take it back so that it should not prejudice a court-martial against him. But Eddie refused. He wanted the court-martial to be perfectly clear about his motives. He had intended to desert and wanted everyone to know it. His reasoning—apparently so harmful to his cause—was absolutely consistent. He wanted to be convicted by a court-martial, even if it meant that he would be sentenced to death. After all, no American soldier in eighty years had actually been executed for desertion. The worst that would happen would be that the death sentence would be commuted to life imprisonment. And when the war was over—perhaps in a year or eighteen months at most—his sentence would be quashed and he would be released. Who was interested in military desertion in peacetime? And the main virtue of this course of action would be that prison was a lot safer a place than the Hürtgen Forest or any frontline position. Certainly prison held no fears for a man like Eddie Slovik, who viewed it as a refuge from a world in which he had failed time after time to measure up to society's standards.

But Eddie Slovik was out of his depth if he thought he could manipulate the actions of the American army. Too many people had fooled themselves into thinking that the German military machine would collapse once the Anglo-American invasion of France had taken place. But they were wrong. The Germans showed a resilience and a determination which was worthy of a better cause. The winter of 1944–5 saw the *Wehrmacht* strike back at the Americans with unprecedented ferocity. Eddie Slovik's division—the 28th—found itself facing annihilation in the vicious Hürtgen Forest fighting. It was no time for American commanders to go soft on shirkers. During the battle of the Bulge many American units fought "backs-to-the-wall" actions to escape destruction. Yet at this time of greatest need senior officers found themselves facing a "self-imposed" manpower crisis. Thousands of soldiers were claiming that they could not face the enemy because of "combat fatigue." Men recruited from civilian life only months before would not tolerate the stress of desperate fighting. As soon as the word got about that there was no real danger of anyone being asked to pay the supreme price for desertion—execution by firing squad—and that the most they could get was a prison sentence, there was a tidal wave of men seeking a way out of the fighting through court-martial and imprisonment. Military law was being challenged so openly that it was tantamount to mutiny. If the law was to be seen to have any teeth, someone would have to be bitten. And that man—that single soldier—was going to be Eddie Slovik. An example was needed—perhaps even a scapegoat—and Eddie's life so far had shown how well fitted he was to play the part. He was one of nature's scapegoats.

So Eddie Slovik was suffering from a misapprehension. During his stay at the divisional stockade at Rott, in Germany, Eddie got to talking to some of the other men who were facing court-martial for desertion. Eddie, it seems, was happy at the time, certain of being taken out of the fighting and sent to the safety of a prison cell. When some men were returning from their trials, Eddie called to them to ask what sentence they had got. The men told him that they had been sentenced to twenty years in prison. But they knew they would never serve such a sentence and seemed almost pleased. Eddie replied, "I'll settle for twenty years right now. How long do you think you'll have to stay in after the war's over?" "Maybe six months," one of the soldiers told him. Eddie's eyes lit up. He had cracked it. Six months in a warm, dry cell, with no Krauts trying to kill him. And afterward, he could return in one piece to Antoinette. It was the best deal he was ever likely to get.

The problem with Eddie Slovik, at least from the point of view of the authorities, was that he could hardly be said to be suffering from "combat fatigue" as he had never even been in combat. He had not panicked after a shell fell close to him, or cracked after a tough spell in the firing line. Eddie Slovik had never even tried to be a soldier. It was the "idea" of fighting that appalled him, rather than the reality. In a more intelligent or cunning man the idea of conscientious objection might have occurred to him as a way out of his trouble. But once Eddie latched on to the idea that a dishonorable discharge and imprisonment was the safest option he clung to it. Even when he was advised to retract his confession he refused, and when he was interviewed by Lieutenant Colonel Henry J. Sumner he declared that the only way he would stay in France would be with a quartermaster's outfit, way back behind the lines. Otherwise he would prefer to be court-martialed. Eddie knew his limitations. Originally classified 4-F by the draft board, he knew that he did not have the makings of a frontline fighting soldier. To reclassify Eddie Slovik as 1-A was a cruel joke. If his country needed him to fight, then his country was in a pretty bad way.

Colonel Sumner had heard many soldiers asking to be taken out of the front line and told Eddie, "We can't transfer everybody out of a rifle company to a warehouse or a supply dump just because they would prefer that sort of soldiering." But Eddie was not listening. He replied to Sumner, "I've made up my mind. I'll take my court-martial."

Eddie Slovik would not have been so confident if he had been able to hear what the senior officer of 109th Infantry—Lieutenant Colonel James Rudder—was saying about the fighting spirit of his men. In his words, "the person that is not willing to fight and die, if need be, for his country has no right to life." The fact was that Rudder and most senior American soldiers believed that the Japanese and the Germans were tougher—man for man—than their American counterparts. The American civilian was too soft. General Patton had issued a warning in 1940: "I'm worried because I'm not sure this country can field a fighting army at this stage in our history. We've pampered and confused our youth. We've talked too much about rights and not enough about duties. Now we've got to try to make them attack and kill. A big percentage of our men won't be worth a goddam to us." Rudder, and the divisional commander, Norman Cota, had had the truth of this hammered into them by the fighting in the Hürtgen Forest and they were not prepared to tolerate any more shirkers. In fact, Cota had already heard all he ever wanted to hear about "combat exhaustion" and psychiatrists finding excuses for men who had not the guts to carry out their duty. If Eddie Slovik was hoping for sympathy he had picked the wrong man in "Dutch" Cota. Having seen his division cut to ribbons in the "green hell" of the Hürtgen Forest, General Cota was unwilling to lose any sleep over a cowardly ex-con like Eddie Slovik. And that was just the problem; once everyone found out about Eddie's previous run-ins with the law they felt he deserved everything that was coming to him. Let him be a scapegoat to save the skins of better men. As Eddie put it himself, "They're shooting me for bread and chewing gum I stole when I was twelve years old."

Private Slovik's trial took place on 11 November, 1944, and lasted just one hundred minutes. Essentially, with Slovik not trying to defend his actions and quite prepared to accept the court's sentence, there was very little to discuss. The findings were that Slovik should be dishonorably discharged and "shot to death with musketry." Eddie's civilian criminal record played no part in the court's decision as it was not made available to the judges. All they saw was an apparently fit young man who was challenging the system and had to be stopped. When he refused to carry out his duties as a soldier and threatened to run away if he were placed in proximity to the enemy, he could have been the spokesman for hundreds of thousand of young Americans who were at that moment locked in a life-or-death struggle with the Germans. To show even the slightest weakness in dealing with Eddie Slovik could have been to court military defeat and national catastrophe. In any case, Eddie's judges knew that the final decision would not be theirs. Many deserters or cowards were sentenced to death by field courts-martial but none of them were ever shot. Eddie's fate rested in the hands of the commander-in-chief, General Dwight D. Eisenhower. All they could do was to respond to the evidence presented to them. And on this evidence alone, Eddie Slovik deserved nothing better than a guilty verdict and a death sentence.

Eddie Slovik was not greatly concerned with the death sentence he had received. It was not unexpected. Someone up the line would commute it to life imprisonment, Eddie supposed, and he was content to be going back to Paris rather than on into Germany. But Eddie's divisional commander, General Norman Cota, was the first man to act in an unexpected fashion. For the first time, Eddie's civilian record became part of the equation. When General Cota's legal adviser, Colonel Somner, became aware of Slovik's "record" it was as if he had suddenly been given an opportunity that had been denied him in other cases. He felt that he found the "scapegoat" who could suffer for everyone else and whom nobody would really miss. As Colonel Somner said, "I never expected Slovik to be shot. Given the common practice up to that time, there was no reason for any of us to think that the Theater Commander would ever actually execute a deserter. But I thought that if ever they wanted a horrible example, this was one. From Slovik's record, the world wasn't going to lose much." Once Slovik's criminal record—slight as it was—became common knowledge, attitudes toward him hardened. Colonel Somner advised General Cota not to recommend clemency in this case and so the death sentence still stood when it reached the theater commander, Dwight Eisenhower. Ideally, Eisenhower would have had time to consider all the facts of the case before reaching a decision. In practice, of course, he was far too busy to even read the letter that Slovik had sent him by way of appeal. The battle of the Bulge was raging and the lives of individuals seemed unimportant in comparison with the progress of the entire war. Eddie Slovik had left it far too late.

To everyone's surprise, Eisenhower confirmed the findings of the court and Eddie Slovik was condemned to die. The staff judge advocate, Brigadier General Edward Betts, had made the following point to the commander-in-chief, which may have proved conclusive. Betts had written:

> He [Slovik] was obstinately determined not to engage in combat, and on two occasions, the second after express warning as to the results, he deserted. He boldly confessed to these offenses and concluded his confession with the statement "so I ran away again AND I'LL RUN AWAY AGAIN IF I HAVE TO GO OUT THEIR" [sic]. There can be no doubt that he deliberately sought the safety and comparative comfort of the guardhouse. To him and to those soldiers who may follow his example, if he achieves his end, confinement is neither deterrent nor punishment. He has directly challenged the authority of the government, and

> future discipline depends upon a resolute reply to this challenge. If the death penalty is ever to be imposed for desertion it should be imposed in this case, not as a punitive measure nor as retribution, but to maintain that discipline upon which alone an army can succeed against the enemy. There was no recommendation for clemency in this case and none is here recommended.[6]

From this point onward Slovik was doomed. Betts had made it clear that sooner or later somebody would have to be shot if the death penalty for desertion was not to become a toothless deterrent. So many American troops had defied the authority of the courts-martial that there was a danger of a complete breakdown in military authority at a time when the Americans were encountering strong opposition from the *Wehrmacht*. When Eddie Slovik later complained that he was being sacrificed for the misdemeanors of his childhood—stealing sweets and bread when he was twelve—he was quite wrong. Eddie Slovik was sacrificed because he had challenged the authority of the American army at the very moment that it was suffering a crisis of confidence. He was shot because he had defied the generals to shoot him. Eisenhower and, at a lower level, Norman Cota could hardly prosecute the war to a successful conclusion if they could be defied by an "ex-con" who would rather stay at home with his wife than fight for his country. In an era of "total war" individuals were expendable, and Eddie Slovik was more expendable than most.

On 31 January, 1945, the execution took place in the courtyard of a house in the village of St. Marie aux Mines. In keeping with the unique nature of the occasion everyone experienced so much nervous tension that the firing squad fired badly. Twelve rifles—one loaded with a blank to allow each man the opportunity to believe that he had not killed the prisoner—were fired from a range of just twenty paces. But even from here they could not kill. At the last moment some of the men must have pulled their shots, leaving it to others to do the dirty work. The atten-

[6]W. B. Huie, *The Execution of Eddie Slovik*, p. 153.

dant physician, Doctor Rougelot, confirmed that none of the bullets had struck the heart. In fact, some bullets struck Eddie Slovik high on the neck and in the left arm, when everyone had been ordered to aim at the heart. The doctor applied his stethoscope to the prisoner's chest and found a faint, rapid heartbeat. Military law was quite relentless in its requirements. As the first volley had failed to kill Eddie Slovik, the firing party was obliged to reload and prepare to fire a second volley. But this was more than the doctor could stand. He knew the prisoner would die within a matter of minutes; more butchery was unnecessary, even barbaric. As Major Fellman ordered the firing party to reload, the chaplain, Father Cummins, swung round and said bitterly, "Give him another volley if you like it so much." The occasion was getting to everyone. The doctor spoke gently to his colleague, "Take it easy, Padre, none of us is enjoying this." But the second volley was not needed. After a few moments more Rougelot pronounced Eddie Slovik dead. Eisenhower had his scapegoat.

In the British army of the First World War, it was assumed that courage was something that could be drilled into the common soldier or developed by immersing him in regimental traditions. Officers, on the other hand, were assumed to be innately brave, courage being something acquired through breeding. The assumption, absurd as it may seem, became almost self-fulfilling, so that boys of eighteen fresh from school displayed courage far beyond their tender years simply because they never entertained the thought for a moment that there was any other way to behave.

The execution of a British officer for cowardice was made especially controversial by the fact that he was a naval officer and, legally, the army authorities had no right to impose the death penalty on him without the permission of the Admiralty. Yet such legal niceties counted for little in the context of the desperate campaign that the British army was conducting on the Somme against the cream of the German army in November 1916. Since the first day of July, the British had been trying to achieve a breakthrough by main force and casualties among both the attacking British and the German defenders had easily passed the million mark. In such a situation, with the ground conditions having deteriorated in the fall rains to those of a vast swamp, watered by rivers of liquid mud, the military authorities felt it was absolutely necessary to make an example of officers as well as men to prevent any wavering. And while the British Grand Fleet was once again in its base at Scapa Flow, with row upon row of great battleships swinging at their anchors never again to fire in anger against the enemy, each one housing a battalion and a half of British servicemen, on the battlefield the British army was fighting for its life in a maelstrom of horror undreamed of by the "Senior Service."

Sublieutenant Edwin Leopold Arthur Dyett was just twenty-one years old when he fell afoul of the military law. He had originally enlisted in the Royal Navy Voluntary Reserve in June 1915, following in the footsteps of his father, who had been a captain in the Merchant Navy. But by 1915 men were being drafted from the navy, which was enormously oversubscribed, to the army which had always been the junior service in Britain. The result was that, contrary to his wishes, Dyett was transferred to the Royal Navy Division—the Nelson Battalion to be exact—which was formed to act as part of the British Expeditionary Force in France. Having seen action in Gallipoli in 1915, the division was sent to the Western Front to take part in the great Somme offensive the following year. The problem was that the division was ill equipped and ill trained to play a role in infantry fighting, and many of its officers longed for service afloat rather than the bitter and unsavory hand-to-hand fighting that prevailed on the Somme. In all probability, Sublieutenant Dyett might have served out the war efficiently in some minor capacity had his wish to serve at sea been granted. Instead, the trepidation that he felt in being a part of a land campaign eventually destroyed him. Yet how much blame should one place on this young officer? From his earliest days in the service it had been noted that he was intensely nervous and liable to succumb to the pressures of a situation. In less desperate times his deficiencies would have been enough to deny him commissioned rank in any of the services.

Dyett's unit was due to take part in an assault against German lines on 13 November, 1916. His reputation as an officer can best be assessed by the fact that he was originally left back at base when the operation was planned but casualties were so heavy that he and another officer were eventually ordered forward as casualty "replacements." Dyett's state of mind immediately prior to the incident that was to cost him his life can best be understood by looking at an extract from a letter he wrote to a friend. Dyett wrote, "There was considerable hostile artillery, gas shells and tear[gas] shells falling all round us, and snipers were all over the place; we had very narrow shaves more than once." Clearly Dyett's nerves were getting the better of him. In heavy fog and with darkness falling, the two officers could not locate their unit. In Dyett's words:

> When it was dark we met a body of men with an officer in charge; they were wanted by Colonel Freyburg, V.C.; there was much confusion and disorder going on and my nerves became strung to the highest extreme. I found that my companion had gone off somewhere with some men. The officer who was leading the party we met was my "one and only enemy", so we were not polite to each other, and as he is junior to me I practically ignored him except for telling him I was going back to [Battalion Headquarters] which I had left an hour or two before in daylight, but finding those places was not as easy a matter as I thought with the result that I got lost for the second time.[7]

As it turned out, Dyett's "one and only enemy" was Sub-lieutenant William Fernie, whose task had been to round up stragglers and return them to the front. Fernie had asked Dyett to help him escort the men he had "rounded up." Dyett had refused. Fernie had placed an unfortunate interpretation on Dyett's refusal, virtually accusing him of being unwilling to accompany him back toward the fight-

[7]A. Babington, *For the Sake of Example,* p. 97.

ing. What Dyett's reason was for refusing to help Fernie we will never know. His frequent references to the state of his nerves leave the impression that Fernie was probably right in guessing that Dyett had lost his nerve. In any case, Dyett now fell in with some more stragglers and together they searched unsuccessfully for their headquarters and finally spent the night in a dugout.

The following morning Dyett reported to brigade HQ, unaware that Fernie's report of his behavior was likely to have serious consequences. In itself, Dyett's alleged misconduct was slight. It hardly compared with the numerous occasions when men had cracked and fled for their lives in the face of the enemy. However, Dyett was an officer and as such could not behave in any way that might be construed as cowardly. In fact, Dyett was his own worst enemy. His continual reference to the state of his nerves, as well as his openness in admitting that he wanted to be transferred from the battlefront to the quieter life of active service at sea, tended to support the view that he was at best fragile in spirit and at worst positively cowardly. Dyett never seemed to be aware that his life was at stake. But forces beyond his knowledge were already at work.

At Dyett's court-martial, the young man chose not to defend himself. Incredibly he was unaware of the risk he was taking by not speaking up for himself. Instead he used a distinguished legal counsel, who confirmed that Dyett had previously asked for a transfer back to the navy, and claimed that he was of such a nervous disposition that he was unsuited to soldiering. As it turned out this was a bad mistake. The court had no brief to show leniency to weak or mollycoddled young men, who would rather be at home with Nanny. Their task was to judge whether or not Edwin Dyett had shirked his duty, and everything they heard seemed to confirm that he had shown cowardice in refusing a direct order. Consequently, he was sentenced to death, though with a recommendation for clemency on the grounds that not only was he young, inexperienced, and of a nervous disposition, but that the day's fighting had been so bad that it was likely to have upset people of a much stronger disposition than Edwin Dyett. These remarks were undoubtedly offered as the strongest possible ground for quashing the death sentence when the case was reviewed. But when the case went to Field Marshal Haig, the sentence was allowed to stand. Haig's view, almost certainly, was that conditions in November 1916 were terrible. They were probably so awful that normal men could not stand them. But this was a war and one had to make sacrifices. Dyett was just such a sacrifice. Many men would have given way if they were shown the slightest opportunity. The line had to be held and discipline however unfair to individuals and however draconian, had to be enforced.

After the war there was much speculation on whether the prosecution of Dyett was, in fact, a political decision. The naval division was unpopular with senior military commanders and was accused of being ill disciplined. In fact, its discipline was more relaxed than was true of many army units, but its members took a fierce pride in its efficiency, which was much resented in higher military circles. Whatever the true reason for prosecuting Dyett, he was found guilty of desertion and cowardice and died before a firing squad. He had been a man thoroughly unsuited to military service and had been punished for weaknesses that could have been predicted during his officer training. Like so many other inadequate soldiers, Dyett suffered for the failings of others.

A Soldier's Morale

Why do some men fight well, individually and as part of a larger unit, when confronted with the awesome threat of battle? Not, surely, because the human race is naturally brave. There is nothing natural about courage. In terms of the dictates of self-preservation, bravery can be distinctly unwise in that it involves danger and a threat to the survival of the individual. Yet history provides us with numerous cases of individual and group courage that are best considered under the heading of "fighting spirit." It is this latter concept that enables human beings to suppress their natural desire for individual safety and overcome the numerous stresses that a military career, notably one pursued in wartime, can impose.

Military thinkers have always recognized the importance of fighting spirit and high morale in soldiers. Xenophon, writing from his experience of warfare in the ancient Persian Empire, accepted that numbers alone were of little value in battle. If an army lacked spirit, its large numbers merely resulted in an aggregation of individual fears. As he said, "No numbers or strength bring victory in war; but whichever army goes into battle stronger in soul, their enemies generally cannot withstand them." Successful generals throughout history have recognized this and have done everything within their power to raise the fighting spirit of their soldiers. Napoleon insisted that the "moral is to the physical as three is to one" and Field Marshal Montgomery rightly claimed that the morale of a soldier was the single greatest factor in warfare. Sometimes small units—or even groups of soldiers—can have an effect in battle out of all proportion to their numbers. Their willpower or fighting spirit alone carries them through apparently impossible conditions. As we will see later, at Gettysburg in 1863, the defense of a decisive hill position by the 20th Maine regiment under Colonel Chamberlain saved the entire Union army and possibly cost the Confederacy victory in the American Civil War. At Balaclava in 1854, the stand by the 93rd Highlanders under Sir Colin Campbell saved Lord Raglan's British army from defeat. Ranking with Chamberlain's defense of Little Round Top must come Major Hankey's desperate counterattack at the first battle of Ypres in 1914 with men from the Worcestershire Regiment, which prevented the Germans turning the Allied flank and reaching the unprotected Channel ports. It is not stretching credulity too far to suggest that this single action may have cost Germany victory in the war in 1914. Certainly, British generals had virtually given up hope. The commander-in-chief, Sir John French, admitted to Haig that he had no reserves to send him. "The Germans have broken us right in," he admitted, "and are pouring through the gap." The BEF was at its last gasp. At this desperate moment, three companies of the Worcestershire Regiment under Major E.B. Hankey, were facing the full onslaught of General von Kluck's First Army. Just 350 officers and men stood between the Germans and the sea. Facing oblivion, Hankey ordered his men to advance toward the village of Gheluvelt, which the Germans had just captured. The British Official Historian described what happened:

> There were still a thousand yards to traverse, and the scene that confronted the Worcestershires was sufficient to demoralize the strongest nerves and shake the finest courage. The stretch of country which they saw in front of them was devoid of cover of any kind; beyond it lay the fences and enclosures of Gheluvelt Chateau and village, in which many houses were in flames. Wounded and stragglers in considerable numbers were making their way back to the shelter of the woods, some of whom cried as the advancing troops passed through them, that to go on was certain death, whilst

> the enemy's high explosive and shrapnel bursting overhead gave point and substance to the warning . . . The first two hundred yards were crossed in one long rush; nevertheless, the Worcestershires were observed by the enemy's artillery directly they appeared in sight, and its fire was redoubled. Over a hundred men fell, but the rest still pressed on. The wire fences of the enclosures near the village and the wall and railing of the chateau grounds were reached and passed, and contact with the enemy's infantry gained. [The Germans] were enjoying the repose of victory, searching for water and looting, and in no expectation of such an onslaught. They offered no organized resistance, and were soon fleeing back in confusion through the village.[8]

Such understatement! It needed the pen of a Napier or a Russell to describe an attack which, in the words of one politician, saved the British Empire. It was far more important than the lunatic antics of Lord Cardigan's Light Brigade at Balaclava. Only Chamberlain's action at Gettysburg was as pregnant with significance. And, above all, what it showed was that with willpower and leadership brave soldiers the world over can achieve almost anything.

Leadership—as provided by Hankey, Chamberlain, or Campbell (above)—needs to be inspirational if it is to achieve anything. As such it can help to make men fight more bravely, more aggressively, or, even in some instances, help them to rediscover the fighting spirit which they had temporarily lost, as one extraordinary example shows. It occurred at St. Quentin in France during the British retreat from Mons in August 1914. British soldiers of the BEF had been retreating for days even though, as far as they were concerned, they had beaten the Germans in two battles, at Mons and Le Cateau. What they failed to appreciate was that with the war being conducted on a previously unimaginable scale, their one hundred thousand men were merely a drop in the ocean compared with the millions of French and Germans soldiers who were involved in the fighting. As a result of the Schlieffen Plan, the German First Army

[8]R. Holmes and J. Keegan, *Soldiers,* p. 41.

under General von Kluck—alone three times as large as the entire BEF—was outflanking the French armies in a great arcing movement and the British were falling back to avoid being trapped. All these strategical considerations meant nothing to men who had been marching for days without sleep and without so much as glimpsing a German. The morale of the British soldiers was very low. They knew they were a match for the Germans but all their officers did was to order them to keep retreating. What sort of war was that? They had had enough of running. They wanted food and sleep, and if the Germans caught them up, then they would surrender.

This was the mood of the British infantry in St. Quentin when a cavalry officer, Major Tom Bridges, happened to ride into the town. What he saw shocked him; it was nothing less than the disintegration of the British army and he, for one, was not going to sit there and watch it happen. Addressing groups of men he tried to encourage them to form up and continue the retreat, but he met with hostility and derision. He was told to keep his "cavalry nose" out of infantry business and that the regiments' colonels had already agreed to surrender to the Germans. Bridges was horrified and went at once to see the mayor of St. Quentin to ask for transport to help move out the wounded. The mayor told Bridges that it was all too late and that the British troops in the town had already agreed to surrender to the Germans. Bridges simply would not believe this until the mayor produced a document signed by two senior British officers agreeing to surrender, which he was going to deliver to the German commander under a white flag.

Bridges told the mayor that as a loyal Frenchman he had no right to surrender to the Germans, but the man explained that he was acting on behalf of the citizens of the town. He had asked the British officers to go and fight the Germans outside the town but they had refused, saying that they had lost their artillery and could not fight. He then asked the British to leave quickly and escape but again the officers had refused, saying the men were too tired. Faced with such blanket refusals, the mayor told Bridges that he had had no alternative but to ask the British to surrender and this the officers had agreed to do.

Bridges took the surrender document, which was signed by Lieutenant Colonel Mainwaring of the Royal Dublin Fusiliers and Lieutenant Colonel Ellington of the 1st Battalion of the Royal Warwickshire Regiment and tucked it into his pocket. He then began organizing transport of all kinds. But the main problem was how to rouse the soldiers and make them see that escape was possible. Bridges came up with a brilliant solution. Lacking a band, he decided to improvise. He went to a toy shop in St. Quentin and bought a toy drum and a pennywhistle. Calling his company trumpeter, he gave the man the whistle and then joined in on the drum, marching round the square and banging the drum to the tune of "The British Grenadiers." Soon the soldiers stood up and began to take notice. Surely the man was mad! But their attitude toward him was already changing. He was providing the leadership that had been missing all day. Within minutes they had begun to form up into a column and, with Bridges and the trumpeter at the head, and with occasional mouth organs joining in, the footsore soldiers marched out of the town. Into the distance they marched with the trumpeter now changing the tune to "It's a long way to Tipperary." The retreat continued—but Bridges had saved the day.

Having recognized the importance of morale, how does one increase it? Circumstances in wartime can be so varied that there is no obvious answer. Sometimes fighting conditions can be deeply depressing in their own right. One has only to think of the fighting at Passchendaele in 1917 in conditions so terrible that when one staff officer from the rear actually saw the front line for the first time he burst into tears at the thought of what he had sent men to fight in. The bitter cold of the Russian front in the Second World War, the blazing heat of the North African desert, the disease-ridden jungles of Burma and New Guinea all tested the morale of European and American troops to the limit and beyond. How did one raise the morale of men starved of food and fighting an enemy who was richly supplied with luxuries? This was the problem for German commanders planning Ludendorff's offensive on the Western Front in 1918. How did you inspire men to fight against an enemy overwhelmingly more numerous, more able, or more savage? The questions accumulate, but the answers are elusive. The problem of fighting spirit is the same in every age. Men must be persuaded to defy their own logic and their own instincts. Sometimes the situation seems suicidal yet the soldier must press forward. He must obey orders to defy his instinct of self-preservation in the interests of a greater purpose. He must set his life and his welfare at naught. He must be prepared to lose all that he has—family, friends, wealth, happiness, and even life—without a second thought. He must obey his officers even at the risk of agonizing death by bayonet, bomb, or bullet. He will experience sights and sounds more dreadful than his darkest dreams and yet overcome his natural wish to run away from such horrors. In doing all these things he will have been a soldier, and he will have displayed fighting spirit to some degree at least. Even the conscript—the civilian shop assistant, bank clerk, or factory worker—must fill the ranks in the tradition of history's warriors. Pen pushers and car mechanics, bus drivers and schoolteachers, must be prepared to see their comrades die and still press on into the storm of enemy fire. They must see sights each day that sear their minds with the horror of war, as was the case with this American captain who witnessed a shell hitting a group of sleeping soldiers:

> At the bottom and down one side of the gully there was a pile of gray shredded fabric. It had no shape and it was not very big. The whole bottom of the gully was coated with a gray powder and you would not have noticed the pile of gray shredded fabric except for a foot and a shoe with no body attached to it. This object lay by the edge of the pile. There was no blood whatever. All the blood had been blown out of the man who wore the shoe.[9]

[9]J. Ellis, *The Sharp End of War,* p. 113.

Ultimately, the individual soldier is so alone in the face of these seemingly elemental forces that he can only find salvation in what has been called "group cohesion and solidarity." He must bond with a small number of other men sharing the same challenge. These will become his comrades in arms. And as these groups bond with other like groups they will form a tight unit—a molecular structure—within which to find a meaning. How these units are grouped and operate becomes the basis for an *esprit de corps,* which hopefully will maximize the fighting potential of the aggregate of individuals. For centuries this "regimental system" has helped to inculcate the military virtues for generations of soldiers in many countries. The regimental "culture" of colors, trophies, emblems, and uniforms has helped to provide individual soldiers with a more substantial "identity" and facilitate their induction into the world of soldiers and war. It has operated as a "club" to which only individuals who have qualified can belong and which offers its members a "home away from home," indeed for many the only real home that they have ever known. With its strict rules of membership the regiment can weed out undesirables and set standards to which its members are proud to adhere and traditional achievements on the battlefield which its recent recruits are eager to replicate.

The suppression of individuality in military training has always had the sensible purpose of ensuring group solidarity and, therefore, group safety. This—the foundation of fighting spirit and not merely individual frenzy—begins in many armies with some form of oath to ritualize the joining process. The German army has always considered this a vital part of a soldier's training. It was always conducted with solemnity akin to a religious service. In pre-1914 days, the individual was made to understand that he was undergoing a transition after which he would never be the same again. One such recruit remembered, "In front of the battalion, drawn up in a square, we had to take the oath of loyalty to the Kaiser. The regimental colors were ceremoniously unfurled and about ten of us, I amongst them, had to step forward. We had to hold up our right hands, had to touch with our left the richly embroidered silken flag and repeat the words of the oath spoken by the adjutant." Oaths of this kind have been common throughout history in armies that possessed the order and discipline of civilized states. Predating even the Roman example were such elite groups as the Theban "Sacred Band," made up of homosexual soldiers and their partners, which was annihilated by Philip of Macedon at Chaeronea in 338 B.C., its members fighting to the last man. Oaths of loyalty to individual chiefs, notably in the Germanic world, were common during the Dark Ages, and the lays and legends of northern Europe, particularly those of the Vikings, record many instances of men dying around their leader and his symbol. The last stand of the English housecarls at Hastings in 1066, fighting around their dead leader Harold Godwineson and his banners of the "Fighting Man" and the "Dragon of Wessex" is one of the most famous examples of this "regimental" loyalty.

The importance of uniformity of action in battle has rarely been allowed to lapse. As a result, new recruits have often been depersonalized, in some cases—one thinks of the American marines or the French Foreign Legion—by haircuts of a fairly radical kind. Fashions are constantly changing. For centuries warriors displayed their manly attributes by the length of their hair. Some recruits in the seventeenth and eighteenth centuries had to grow their hair longer or style it in a particular way to become an accepted part of their regiment. Once the hair had been shorn or restyled, attention passed to the uniform. This was the most clear indication of a soldier's regimental identity. In Western warfare uniforms did not develop until the mid seventeenth century, with the progressive Swedish king Gustavus Adolphus in the forefront of developments. During the Thirty Years' War—a war of mercenaries if there ever was one—men of the same nationality and even of the same religion could be fighting on opposite sides. Some reliable method of identification was necessary: battle emblems—colored sashes, flowers, or greenery—were too unreliable. They could easily be removed and replaced as the situation demanded. Sir Thomas Fairfax, the Parliamentary general, was able to ride unmolested through much of the Royalist army at the battle of Marston Moor in 1644 by the simple action of removing the white sash the Parliamentarians were using for identification. And so developed the regimental uniform in all its splendor. Initially designed for identification, it soon took on far greater significance. Rather than the defensive purpose of preventing its wearer being attacked by his own colleagues, it began to evolve as an aggressive feature of its wearer. In most cases it was cut so as to make its wearer either larger—broader or taller—or more fearful. Its show of finery exuded confidence. It spoke for the individual within. If its regiment had won numerous battle honors, then its wearer was to be feared as belonging to a formidable force. Whether one wishes to equate developments in uniform design to the display principle of the animal world as some historians have suggested, ie. that "one threatens by making oneself bigger—whether by raising one's hackles, wearing combs in one's hair, or putting on a bearskin" is a moot point. Certainly the "Death's head Hussars," with their skull emblems, may have thought so. Ultimately, however, bullets are unimpressed. Artillery shells are no respecters of regimental finery.

The main contribution to fighting spirit that a regiment can make is to drill or train the soldier to a point where his own efficiency is beneficial both to himself and to his colleagues. A recruit's morale is boosted by his apparent mastery of his military environment. Long hours spent on the parade ground, drilling and "square-bashing," is fundamental to any soldier's development. There is nothing natural about the soldier's life. Even what natural aggression humans have will have been drawn from the recruit by his cultural and educational experience. Societies in peacetime have needed to channel aggression into more acceptable forms, particularly competitive sports. The British love of sport—no nineteenth century creation but apparent in various forms from medieval times—may almost be seen as an alternative to the militarism of neighboring states. The huge standing armies of European states from the eighteenth century onward was one way of channeling the natural energies of their young men. Conscription and military training in France and Germany enabled the most rebellious members of the population to be controlled and regulated, while still providing a service to the community. Such an absence in Britain and the United States may be reflected in crime figures—on the one hand—and in the number of young men involved in sporting activities as participants or spectators. It is significant that many of the most important modern world sports developed in Britain or the United States during the nineteenth century. German commanders have traditionally criticized the British approach to war as being no different from their approach to sport: the British had retaliated by likening the German attitude to sport as akin to their approach to war. Nevertheless, the essentially "amateur" approach to war on the part of the British can be exaggerated. Significantly, in 1914, the British Expeditionary Force sent to France was infinitely more efficient, man-for-man, than any of the German units it encountered. There was nothing amateur about the "fifteen aimed shots a minute" that slaughtered the Germans at Mons and First Ypres, where the astonished Germans claimed they were facing massed machine guns rather than the single line of riflemen who in fact held the line. The "Old Contemptibles" were an elite force like Wellington's Peninsular veterans, who possessed a clear belief in their superiority that was justified by the results.

The aim of the eighteenth century Prussian drillmasters was to ensure that the men in their charge could master the weapons of the age with a precision that set them above the men of Prussia's enemies, of whom there were many. The quality of the recruits in the warring states of Frederick the Great's days varied little, nor indeed did their weapons. It can even be argued that the Prussian musket was less efficient than the French or Austrian weapon. How then did the Prussians win so many of their battles, against more numerous and even better-armed adversaries? It would be simple to point to the leadership provided, by Frederick and by numerous others, but this would be less accurate than some critics might suppose. After the opening volleys in a battle of the time, even Frederick himself admitted there was little he could do to secure a successful outcome.

That was dependent on the raw material, the soldiers who did the fighting, and therefore to the drillmasters, the men who made the soldiers what they were.

Confidence on the battlefield—notably during the eighteenth and nineteenth centuries—came from proficiency in drill. Battles were often almost "bloody drills" with units maneuvering according to well-established patterns. In this case a small advantage in speed of loading or unity of volley fire could make all the difference to the outcome of an entire battle. The superiority of British firepower at Fontenoy against the French was marked and by 1759 the advantage had grown so that the British infantry were able to shatter both French infantry and cavalry at Minden. Yet—as we will see—the confidence the British enjoyed against European opponents could disappear in a moment when the drills to which they were accustomed were no longer appropriate. The work of the drillmasters, therefore, could only be successful within the confines of the European battlefield. Presented with a different challenge—Indians or irregulars, like Highlanders, for example—the fighting spirit so laboriously cultivated on the parade ground could prove ephemeral.

Drill has not only been used to suppress individuality of thought and promote mindless uniformity of action. It can also provide a useful prop for a soldier whose circumstances make careful thought impossible. Men suffering from combat exhaustion are hardly likely to make sound or original decisions based on a careful analysis of the situation in which they find themselves. Where environmental and personal pressures are enormous through heat or cold, mud, sand, or snow, disease or injury, it is reassuring to be able to fall back on something acquired almost like a second skin. As one GI observed from his own experience during the Second World War:

> In mortal danger, numerous soldiers enter into a dazed condition in which all sharpness of consciousness is lost. When in this stage . . . they can function like cells in a military organism, doing what is expected of them because it has become automatic. It is astonishing how much of the business of war can still be carried on by men who act as

> automatons, behaving almost as mechanically as the machine they operate.[10]

Ultimately drill achieves shared experience—even shared suffering serves to draw individuals together—and through this a feeling of "belonging" to something beyond the individual. A recruit thus becomes linked with everyone who currently wears the regimental uniform, and also all those who ever wore the uniform. As the Comte de Guibert, an eighteenth century French military thinker observed, *esprit de corps* was something to be cultivated:

> Personal bravery of a single individual does not decide on the day of battle, but the bravery of the unit, and the latter rests on the good opinion and the confidence that each individual places in the unit to which he belongs. The exterior splendour, the regularity of movements, the adroitness and at the same time the firmness of the mass—all this gives the individual soldier the safe and calming conviction that nothing can withstand his particular regiment or battalion.[11]

As I hope to have shown in "Heroes and Villains," the role of the individual in battle can exceed that accorded to it by an eighteenth century mind like Guibert's. Nevertheless, his description of the confidence accorded to individuals by membership of a group is entirely true. And it is this group identity that can make him stand fast when fear is all around him and he would rather run away. Regimental histories do not record the men who cracked and ran, only the ones who stood and fought. The recruit will therefore be surrounded entirely by success, by honor, and by glory. According to the histories, even those who perished fell in a great cause, surrounded by mounds of enemy dead, and passing on their fighting spirit and their own noble natures to those who would come after them and wear the regimental uniform. There is no room for cyni-

---

[10]R. Holmes and J. Keegan, *Soldiers,* p. 44.

[11]Ibid., p. 46.

cism in regimental histories, which are normally written precisely by the men who most strongly feel the *esprit de corps* that the regiment gave them. Those who could have redressed the balance, spoken of the cowardice and the folly of earlier generations have no place in the sacred pages.

Regimental spirit might enable a well-dressed subaltern to "cut a dash" in the London clubs or even impress the ladies, but it was on the battlefield that it became metamorphosed into something more useful, a feeling of pride and certainty that removed doubts about personal courage and the threat of defeat and death. Regimental battle honors could be called out almost like magic words which inspired the men to redouble their efforts at difficult moments. General Slim remembers such an occasion during the First World War. A battalion of the Warwickshire Regiment was faltering during an action when a private soldier suddenly yelled out, "Heads up the Warwicks! Show the blighters your cap badges!" At this the men seemed to rally. More famous examples abound—like the immortal dying words of Colonel Inglis at Albuera in 1811: "Die hard, 57th, die hard!" Generations of soldiers from this famous regiment—later the Middlesex Regiment—have been proud to be known as "The Diehards."

In these cynical days it is easy to overlook the significance of patriotism, yet if we are to understand why men fought well in battle, it is necessary to examine this aspect of human motivation. Just as *esprit de corps* has a great significance for a fighting man, by providing a group identity, so is membership in another exclusive club, to which he belongs by birth: the nation. Again traditions can inspire him to fight: old triumphs or perhaps even old injustices. During the Falklands War, for example, it was difficult for hard-bitten British professional soldiers to take seriously the commitment of young Argentinian recruits to the concept of the gloomy, windswept Falklands as the "Malvinas," a historical part of Argentina, which they would willingly sacrifice their lives to regain. Clearly, indoctrination and propaganda had played their parts in this, and cynical politicians had used the Malvinas issue to earn a spurious popularity. It is difficult to avoid agreeing with Samuel Johnson that there are occasions when "Patriotism is the last refuge of a scoundrel." Yet it would be wrong to overlook the effect that "the blue line of the Vosges"—regaining Alsace and Lorraine from Germany—had on millions of French *poilus* in the First World War, or the jingoism that inspired British troops in the Crimean War against Russia.

Patriotism was a significant weapon of the recruiting officer in 1914, particularly in Britain but also in Germany and elsewhere. It should not be assumed that the German army of 1914 was entirely a conscripted one. Hundreds of thousands of young Germans—the majority university students—joined up at the first opportunity, inspired by a love of their country. Whether their patriotism stayed with them for long at the front was more a personal issue, though it undoubtedly inspired many to fight for their "fatherland" and all that they thought it stood for. Readers in the English-speaking world may be surprised to find how much the German experience mirrored that of the first British volunteers in 1914. German soldiers were well educated and yet they had a strong sense of duty. They could justify any amount of severity on the grounds that it "must be the done for the sake of the Fatherland." In Britain, it must be remembered, virtually all her fighting soldiers until 1916 were volunteers. Their fighting spirit was sustained by at least a veneer of patriotism. Beneath this surface there were as many personal reasons for fighting as there were men, yet the patriotism was very real for all of them. On the outbreak of war in 1914, one young man, W.T. Colyer, not long out of school, spoke for many in England: "My bosom swelled and I clenched my fist. I wished to goodness I were in the army. I felt restless, excited, eager to do something for the cause of England." George Morgan, a working-class lad who joined the 1st "Bradford Pals"—part of the West Yorks Regiment—was just as patriotic: "We had been brought up to believe that Britain was the best country in the world and we wanted to defend her. The history taught us at school showed that we were better than other people (didn't we always win the last war?) and now all the news was that Germany was the aggressor and we wanted to show the Germans what we could do." Some of the upper-class volunteers—and it must be remembered that the great English public [private] schools turned out their senior boys to be officers on an incredible scale—had a simplistic, sentimental approach to the coming war that they were only to lose when the sordid reality of the fighting asserted itself in 1915. A young lieutenant named Carver summed up this romantic, Rupert Brooke-style image of the mood in 1914: "I always feel that I am fighting for England—English fields, lanes, trees, good days in England, all that is synonymous with liberty." What may sound remarkably naive to the modern reader was very real to those millions who passed the numerous posters of Lord Kitchener, the new war minister, pointing his finger at every Englishman and telling him "Your country needs you." Kitchener symbolized not only England itself but the Victorian father figure, whom most young men of the time found it difficult to disobey. As C.E. Montague wrote in his book *Disenchantment,* "Each of the volunteers seriously thought of himself as a molecule in the body of a nation that was really, and not just figuratively, 'straining every nerve' to discharge an obligation of honour . . . All the air was ringing with rousing assurances. France to be saved, Belgium righted, freedom and civilization rewon . . ." And all this could only be done by the English. A simple British soldier said as much: "I felt that what we were going to do was something that had just got to be done. Had not the Kaiser invaded Belgium and were not the Germans a bad crowd?" And so millions of Britons set off for the front as self-appointed "crusaders," a role that was to be taken from them in 1918 by the eager and naive young Americans who arrived in France.

But the problem of patriotism as a motivating force was simple; it might aid recruitment and help man the front line, but could it, of itself, make men fight better? As an ideal it tended to die as disillusionment set in. Men motivated by patriotism alone were few. After periods in action more fundamental factors began to assert themselves. A generation later a captured German officer revealed as much when questioned about the political attitudes of his men: "When you ask such a question, I realise well that you have no idea of what makes a soldier fight. The soldiers lie in their holes and are happy if they live through the next day. If we think at all, it's about the end of the war and then home."

However, self-preservation as a motive for fighting is too obvious and unhelpful a view. In most cases self-preservation is best achieved by not fighting at all—by running away, by hiding, or by surrendering at the first opportunity. The fact that most soldiers do not choose this way out indicates that they retain a sense of duty, if not to their country at least to their units or their friends and, most of all, to themselves. Modern soldiers frequently explain their motivation in more sophisticated terms than would have been the case a hundred or so years ago. A Crimean War veteran would have felt it at least unmanly and at most downright dishonorable to admit that he was motivated in battle by a desire to preserve his own life, particularly if this required him to let down others. Not every man shared Falstaff's cynical view of honor:

> Honour pricks me on. Yea, but how if honour pricks me off when I am come on? how then? Can honour set-to a leg? No. Or an arm? No. Or take away the grief of a wound? No. Honour hath no skill in surgery then? No. What is honour? A word. What is that word, honour? Air. A trim reckoning. Who hath it? He that died o' Wednesday.[12]

Honor might seem to be just a word to some, but its root is very much in the obligations we feel to others. It is linked with fear of failure, not merely in the pursuit of a task or goal, but in our duty to others. Nobody feels happy who lacks self-worth and this is generally based on our perception of how others see us. Pursuit of honor can also help the soldier maintain his masculine image by controlling his fears or tendency to flight. In the words of Richard Holmes, "Most men have more physical courage than they

[12]William Shakespeare, *Henry IV,* Part One, Act V, Scene 1.

do moral courage, and regard the possibility of death or injury with less terror than they do the probability of disgrace." As a result, the phrase "Better death than dishonor" is not mere melodrama but contains an elemental truth for men in wartime. After all, one does not have to live with death, but one does—presumably—with dishonor. That astute thinker S.L.A. Marshall concluded from his researches into American military history, "Personal honor is valued more than life itself by the majority of men." German soldiers of the *Wehrmacht* were much criticized for fighting so savagely in the last months of the war, when defeat was certain and the awful truths of the regime they represented were becoming known even to the field soldiers. Surely desertion was the only course for decent human beings. General Bradley reported one German general as explaining his unwillingness to surrender: "I am a professional and I obey my orders." He was not a mere automaton of the kind who manned Prussian armies in the eighteenth century. He had standards below which he could not maintain his self-respect, and disobedience to orders and dereliction of duty were anathema to him. Perhaps he failed in his duty to the human race, but he did not fail in his duty to himself and his colleagues. A similar attitude impelled a young Bavarian soldier to prefer death to captivity and loss of self-esteem, when a Bavarian position was overrun by Canadian soldiers in 1918. One of the Canadians, Jack Seely, recorded the Bavarian's last moments:

> Hundreds of them were shot . . . Hundreds more stood their ground and were shot at point-blank range or were killed with the bayonet. Not one single man surrendered. as I rode through the wood . . . I saw a handsome young Bavarian twenty yards in front of me miss an approaching Strathcona, and, as a consequence, receive a bayonet thrust right through the neck. He sank down with his back against a tree, the blood pouring from his throat. As I came closer up to him I shouted in German, "Lie still, a stretcher-bearer will look after you." His eyes in his ashen-grey face seemed to blaze fire as he snatched up his rifle and fired his last

> shot at me, saying loudly: "No, no, I will not die a prisoner." Then he collapsed in a heap.[13]

What of the soldiers who fight for pay? For many years Britain employed troops from the Indian subcontinent in a mercenary capacity. However, men like the Gurkhas needed no financial incentive to inspire them with fighting spirit. These men joined the army as an honorable profession and fought with high morale to preserve their own and their unit's honor. As far back as 1858, in the aftermath of the Indian Mutiny, Sir John Lawrence, got to the heart of the question of pay when he said, "There can be no question that a contented servant is better than a discontented one. Unfortunately we have tried too much to purchase the contentment of our Native armies by increased pay. This has not answered its purpose. It has enriched the sepoys but not satisfied them." Money was not enough. Mercenaries have always fought for pay, but their allegiance has been questionable. Where they fought for their honor and their professional pride their performance was generally better. The sixteenth century Swiss mercenaries were formidable fighters. But so "professional" was their attitude that it occasionally clashed with loyalty to their paymaster. Where there was no "honor," such men could always be "bought off" if the enemy offered them more money.

Recruits into the Indian army shared the many motives that any young men have when they "join up" for professional reasons. However, in India the military profession was regarded as a more "honorable" one than in the Anglo-Saxon world. There were specific reasons that increased the importance of valor in action. The most important motive for joining up among Indians was the hope of winning honor and fame by taking part in battle and performing brave deeds. The Urdu word *Izzat* may be translated as "honor or reputation, credit or prestige." Men gained or lost *Izzat* in their own eyes and those of their

[13]Richard Holmes, *Firing Line,* p. 306.

comrades according to a widely understood code of conduct. Fame was closely linked to *Izzat.* Those who added to their *Izzat* might be remembered and spoken of even after their deaths. To die on the battlefield was, for the Indian sepoy, to achieve a glorious end. In India, the British authorities were aware of the power of *Izzat,* and tried to structure rewards to their Indian soldiers accordingly. In 1837 two orders of merit were created to reward Indian soldiers, along with titles such as *Bahadur,* meaning "chief or brave one." During the First World War, the British found that the Indian soldiers were impressed by awards of medals and titles to courageous sepoys, along with appropriate ceremonies. Indian soldiers were greatly inspired by the award of medals, particularly if they came from the king/emperor himself. Sepoys regarded "honor" as something won or lost by individuals on their own behalf, though *Izzat* could apply to a corporate body of which they were part, like a regiment or a clan. Even India itself could gain honor from an individual's action. "Our King will conquer his enemies very soon," wrote a Jat lancer. "He will say, 'My brave Sepoys have done splendidly and have gained great fame for India.' " Usually, however, benefits would apply to clans or castes. As one Jat cavalryman wrote, "If we get any chance we will show the stuff our caste is made of." But just as honor could be gained by glorious deeds, so could shame. One Indian Muslim asked his son to gain leave to return to India, but his son replied, "It is sad to think that you served for twenty-five years and yet you are unable to understand the present state of affairs! Moreover, you say all this on a postcard and the consequence is that people ridicule me for what you say. Give my compliments to the writer of the card! Is it possible that anyone could go home on leave at such a time as this? Never write me such a letter again." To the sepoys, personal honor was more important even than family or friends.

Indian perceptions of Western warfare were important to understand if Indian units were to be inspired to achieve high morale. One man explained his impressions on watching Indian troops advancing during the battle of the Somme in 1916. How poetic it is compared with the naturalistic descriptions of British and German observers in the section titled "Fight or Flight." What is most apparent, however, is the joy experienced by the man watching, not cynically expressed by one free of the danger but exulting aesthetically in the whole military process:

> One forgets the achievements of Bonaparte when one sees what our men have done. How our heroes have gone forward, quite regardless of life, and crushed the head of the enemy on the ground! Battalion after battalion follow their music, filled with enthusiasm, just as a snake dances to the pipe of the charmer and darts forward to strike. Battalions go forward with even step, steadily and firmly, just as an elephant moves along the road swaying slightly from side to side, to show the worth of their valour. Truly, even from the enemy's lips they must have wrung applause. Even the heavens do not cease shedding tears on our warriors, so great is their valour.[14]

Of all the European powers, Britain achieved a better fighting spirit from her colonial forces than any other. This was a result of understanding the mentality of the Indian troops particularly and realizing the importance of personal and group honor to the sepoy soldiers.

It is as well for military historians, as well as soldiers, not to seek solutions only in that which is most worthy and laudable. Certainly, men will often claim the highest of motives for themselves. Tradition, loyalty, honor, duty are ideals worth fighting for; but as the history of warfare shows, ideals take a terrible battering in wartime, and it is only the most remarkable of soldiers who can engage in the day-to-day struggle and retain his ideals. Sometimes men need more obvious, more tangible targets. Loot or booty have generally been the target of the common soldier. Where kings and emperors won land and fame, and their officers glory and promotion, for the footslogger it was more a question of something to soothe the wounds after the battle is over; something tangible to put by for

[14]D. Omissi, *The Sepoy and the Raj,* p. 79.

an uncertain old age, possibly crippled by amputation and otherwise forced to beg. Such rewards—a ring here, a purse there—may well have been at the root of fighting spirit from the time that men first began to fight among themselves. Ransoms were generally reserved for officers. For the lowly soldiers it was more a question of pillaging the enemy's camp or cutting purses from the dead or stripping their corpses. It is surprising to find the search for loot as a motive of the German soldiers during the March Offensive of 1918. There can be no doubt that when food has become a luxury, as in this case, the fighting spirit of the soldiers will be much enhanced by the prospect of a change of diet. Otherwise, unless the prospect of pillage is a particularly attractive one, the attraction of loot is more opportunistic and less motivational.

However, few soldiers—and the British have been some of the worst offenders—have been able to resist the lure of liquor. Whether alcohol before a battle improves a man's fighting spirit is questionable. It may certainly dull his fears; it may even in a few cases make him "roaring drunk"; yet most armies have provided liquor to their men to try to boost morale. The issue of rum before British soldiers went "over the top" was a common habit during the First World War. While it may have settled the nerves, it could do little for the man facing the basic reality of "kill or be killed," which had its own remarkably effective way of sobering any soldier. Ironically, the British often behaved as madly and rashly as drunkards when they were most sober. The Russians were convinced that the Light Brigade could only have undertaken its mad ride because all the troopers were drunk. In fact, as we know, they were merely brave men led by idiots, or "lions led by donkeys." Some of them, meeting their commander, Lord Cardigan, riding back from the Russian lines, asked him if they could go again. They must have viewed it like some thrilling ride on a "roller coaster" rather than the "wall of death" it so obviously was. Nevertheless, history records many examples where strong spirits or special drugs were used to help men fight bravely in battle. John Keegan and Richard Holmes tell us that the Vikings used a "dried fungus whose hallucogenic effects blurred the images of battle." These drugged warriors were known as berserkers and were reputed to fight with the strength of several men. So fierce were they, that they only looked for enemies at the front and they were usually followed by an ordinary soldier to guard their vulnerable backs. Whether or not this really indicates fighting spirit—as against madness, drunkenness, or frenzy—is a moot point. Certainly, the heroism and fighting spirit of the Viking who held the bridge against Harold Godwineson's housecarls at the battle of Stamford Bridge in 1066, was undoubted. He may have been a berserker, because he eventually fell to a "low blow," delivered from beneath by a Saxon soldier in a coracle. Soldiers will always drink to relieve the misery and boredom of their lives, and even when alcohol is not issued officially, soldiers will bolster themselves for the task ahead with a few drinks. At Waterloo in 1815, one of the most noteworthy incidents involved the bravery of Corporal John Shaw of the Life Guards, a renowned bare-knuckle fighter and a determined carouser. During a hand-to-hand struggle between the rival cavalry, Shaw fought an extraordinary battle against his opponents, killing nine of them, hacking them down right and left and then hitting them with his helmet and even punching and strangling them after his saber had broken. Shaw had been drinking heavily before the battle and may well have been inspired to a fury by the strong drink. Perhaps one should leave the final word to the medical officer of the Black Watch in 1922, who observed, "had it not been for the rum ration I do not think we should have won the war."

As in the treatment of all human conditions—and the cultivation of fighting spirit is a mental process that is vital to a soldier's health—action can be taken that is either preventive or curative. So far, we have been looking at ways of dealing with the healthy soldier, whose morale needs to be maximized to help him operate efficiently. However, sometimes this is not enough. The soldier may have succumbed to the physical and mental stresses of his vocation and thus become susceptible to combat exhaustion and nervous collapse. Therefore, every effort must be made to minimize the damaging effects of service life in order to achieve efficiency, despite the difficulties of wartime.

The truth is that a soldier's life in wartime will involve him in situations, not all of them dangerous to his physical survival, which are damaging to his mental equilibrium. His fighting spirit will hardly survive intact if he is subject to more stress than his constitution can take. The British army in the Crimea might be thought to disprove this (*see* "Fight or Flight") yet if consideration is given to the living conditions of the common folk in Britain at the time, from whom the majority of soldiers were drawn, and their relatively low expectations from life, then it may be seen that the conditions they endured in the Crimea were less depressing for them than might be suspected by modern historians. The greater problems are experienced by modern armies raised, as these have been in Britain and the United States, from an essentially civilian society, unused to if not actually hostile to the requirements of the military services. Men accustomed to high living standards and a high comfort level in their civilian lives resent the harsher conditions to which army life condemns them. Unable to remedy their situations by any personal action, they find themselves depressed if not actually feeling oppressed, a situation hardly conducive to positive thinking, which is at the root of good morale.

Research into the experiences of the American army, notably during World War Two, has shown that the fighting spirit of an individual soldier is the product of his capacity to adapt to his changed circumstances within a military environment. No amount of *esprit de corps* can completely overcome the individual's feeling of alienation from his civilian lifestyle. This feeling of loss can at best be minimized, but it can rarely be totally removed. In this sense, the fighting spirit of an individual or unit will be variable and susceptible to sudden collapse should links with the civilian past be totally lost.

A military combat situation is probably the most stressful situation in which any human being can find himself. The simple imperative of "kill or be killed" is as fundamental as anything in life. Fighting spirit depends upon the ability of the individual or the group to rationalize the stresses and so resist them, at least for a while. Eventual breakdown is inevitable, and, certainly in the last fifty years or so, this has resulted in units being removed from the front line for periods of rest and recuperation, which are imperative in maintaining morale and a healthy state of mind.

The stresses of war are quite beyond the normal experiences of a civilian conscript and require not only an adjustment in his lifestyle but also an adjustment in his perceptions. Fundamentally he will never before have been placed in a situation where he is required to kill another human being. For many civilian soldiers this remains the greatest stress of all. In the heat of the moment, in a melee of hand-to-hand fighting, moral imperatives may be lifted temporarily, but before or after a battle the act of killing remains a culturally enforced anathema.

> The Army cannot unmake [Western man] . . . It must reckon with the fact that he comes from a civilization in which aggression, connected with the taking of life, is prohibited and unacceptable. The teaching and ideals of this civilization are against killing, against taking advantage. The fear of aggression has been expressed to him so strongly and absorbed by him so deeply and pervadingly—practically with his mother's milk—that it is part of the normal man's make-up. This is his greatest handicap when he enters combat. It stays his trigger-finger though he is hardly conscious that it is a restraint in him.[15]

Many soldiers fail to cope with the conflict of values that occurs in wartime. In civilian life he may always have been prepared to "mouth" the words of the national anthem, express a certain loyalty for his homeland and his national representatives in sporting events; he might even have bought his own nation's products instead of foreign imports, yet this has been a pale form of nationalism. He had

[15]J. Costello, *Love, Sex and War,* p. 136.

never really thought that he would ever be expected to lay down his life for his country. For his family and friends he may well have been prepared to make sacrifices, even—if the ultimate sacrifice was demanded of him—of his life. But once the soldier has been conscripted and is approaching active service, he suddenly realizes that there is an inevitable conflict in his situation. Internally he is a dutiful son or husband or father. He feels that he must do his best for those he loves and who are dependent on him. This cannot involve risking his life or health on a battlefield. If he is killed or crippled, who will look after his loved ones? Their distress—and perhaps even financial hardship—will be his fault. He risked his life when he had no right to, at least in family terms. But society at large is not tuned in to these private thoughts. As an American—or Briton, or Frenchman—he owes a higher obligation to his country. This obligation—legally enforceable in a way that familial duty never is—can be so strong that it requires of him many things that will make him and his family unhappy, damage his health and his economic prospects, and, ultimately, present him with the stark choice of killing or being killed himself. The state can order him to murder or be murdered by his fellow human beings. It can challenge his most profound religious beliefs and punish him with imprisonment and even death if he fails to be what he may never have wanted to be—a combat soldier. Where his family would always tell him to take care, the state may demand of him the very opposite. His inclination to take cover in the face of danger is unlikely to be the response most valued by his military commanders. A caring man may thus become the target for a critical—even abusive—response from his sergeant for shirking his duty. The stress imposed by the cultural clash between civilian and military mores can bring about nervous collapse. Although this is a recent discovery, it must have been a factor in most armies of the last two hundred years. There is no simple way out, except perhaps to reassure the individual soldier that there is a higher moral duty involved in wartime than he has encountered through his religious or ethical education in time of peace. This may or may not prove effective.

Where it fails, the soldier is likely to choose the path of "conscientious objection."

The physical and mental health of young soldiers has been assumed to include the need for a sexual element. What had been possible for the young men in peacetime through girlfriends and wives may become completely impossible in wartime conditions. Psychologists have shown that the sexual act involves far more than mere physical gratification. According to Stouffer, in *The American Soldier,* sexual activity represents an individual's search for

> ... security, for feeling oneself a valued person, for reassurance that one is considered worth affection. . . . Under great anxiety and insecurity, men tended to lose many of their usual long-term perspectives. At the same time, their need for emotional reassurance was especially great; faced with the immediate possibility of personal annihilation amid the vast impersonal destruction of war, hedonistic drives and socially derived needs combined to make sexual deprivation a major stress.[16]

There is nothing new in this. Deprived of sexual gratification, soldiers the world over have become depressed. The need for outlets—normally through brothels though sometimes through a compliant local population—has long been recognized in progressive armies. According to John Costello:

> The men in a successfully trained army or navy are stamped into a mould. Their barrack-talk becomes typical, for soldiers are taught in a harsh and brutal school. They cannot, they must not, be mollycoddled, and this very education befits nature, induces sexual aggression, and makes them the stern, dynamic type we associate with men of the armed forces. This sexual aggressiveness cannot be stifled . . . This very sexual drive is amplified because of fresh air, good food, and exercise, and exaggerated by the salacious barracks talk. It cannot be sublimated by hard work or the soft whinings of Victorian minds. How important this li-

[16]S. Stouffer, *The American Soldier, Volume 2,* p. 80.

bido was considered historically can be gathered from the words of Gian Maria, Duke of Milan, who after his defeat stated, "My men had ceased to speak of women, I knew I was beaten."[17]

Combined with sexual deprivation as a cause of low morale comes the loss of personal privacy. Twentieth century civilians, particularly those in the more urbanized and industrialized countries, have learned the value of being alone. This has as its basis the need for self-respect. It does not reflect a lack of affection for friends, family, or other loved ones, merely that the individual needs to function individually on occasions rather than as part of a group. Military life is absolutely hostile to this human need, and, where the need is very great in an individual, low morale is guaranteed if the need is treated with the sort of crass hostility that we encounter in the following examples. This soldier found the process of military service dehumanizing:

> Living in herds and schools like steers or fish, where men (suddenly missing deeply the wives or girlfriends they left so adventurously two weeks before) literally could not find the privacy to masturbate even in the latrines. Being laughed at, insulted, upbraided, held up to ridicule, and fed like pigs at a trough with absolutely no recourse or rights to uphold their treasured individuality before any parent, lover, teacher or tribune. Harassed to rise at five in the morning, harassed to be in bed at nine-thirty at night.[18]

Sensitivity in a soldier is hardly a trait worth developing, at least in the eyes of any drillmaster. However, the stresses involved in "toilet taboos" could strike surprisingly deep in soldiers unsure of their sexual persuasion.

---

[17]J. Costello, *Love, Sex and War*, p. 120.
[18]Ibid., p. 119.

> Toilet taboos were suspended for the duration. Fifty of us shared one latrine and took turns at cleaning it, in a symphony of grunts and smells and flushing noises. There were no doors on the booths, nor privacy at the urinal. Answering nature's call meant subjecting yourself to loud and detailed criticism—perceptive and merciless descriptions of your sex organs, ranging from glowing admiration; brilliant critiques of your style of defecation, with learned footnotes on gas-passing. Expert discussions gave new meaning to your technique of urination—which hand, how many, or no hands at all or how nonchalant you managed to look. We soon learned to flaunt our genitals and brag about our toilet mannerisms. Anyone who was modest about these was immediately and forever labelled a homosexual.[19]

The individual could not fight the system. Either he learned to suspend his sensitivity and desire for solitude or he cracked up. Military authorities have often wanted civilian life more to resemble military life: one has only to think of eighteenth century Prussia or totalitarian Germany, but—one suspends a verdict on military systems in Scandinavia—so far nobody has suggested that military life should more resemble civilian life. However one "civilizes" army training, one must never forget that, in the final analysis, the task of a soldier is what it always has been, "to kill or be killed." The weapons may have changed, but the prime directive remains the same. In which case, one is on dangerous ground in mollycoddling recruits. Clearly, the problem is more one of personnel selection than of training. Some individuals are so unsuited to military life that they should not be forced to conform to its standards.

It is relatively easy to see why many civilian soldiers have developed low morale and poor fighting spirit during their period of service. They have cracked—through nervous exhaustion and stress—or played the coward by running away in battle or deserting from their unit. The

[19]J. Costello, *Love, Sex and War,* p. 119.

application of the term "coward" to such men raises several points; not least is whether or not they have shown the greater courage in running away than in conforming with "the herd" in actions of which they violently disapprove. Thomas Fuller wrote in 1732, "Many would be cowards if they had courage enough." And in Fuller's day it took great courage, for death and dishonor awaited the coward if he were captured, and so it remained in Britain up until 1920.

But men have always struggled against the horrors of warfare, the physical discomforts of their lifestyle, disease, squalor, and boredom, the shortages of proper clothing and proper food, the fatigue and lack of sleep, the isolation from their families and their homes, the restrictions on their freedom of movement, the rejection of their values, the ridicule of their personal characteristics, their treatment as "things" rather than people, as units or numbers in some great board games played by men who seemed to care nothing for their welfare or for their shared humanity. All of these factors and hundreds more like them can contribute to lowered morale and poor fighting spirit. Listed like this they seem an indictment of war, which, in a way they are. However, while war remains a tool of national politics, the state will retain the right to employ its citizens in this way. Probably the greatest clash in values occurred in Britain in 1914, where, for the first time, Britain found herself with a massed civilian army. Educated and sophisticated recruits placed themselves voluntarily in the hands of military authorities who had never had to deal with such people before. Previously, soldiers had come from the working classes only and had usually been men hardened by a life of toil and poverty. Such men did not cherish solitude or claim privacy; they were content to take orders and do their duty. Yet the volunteers of 1914 represented the "cream" of the country. Never before had any army been drawn from the highest social, intellectual, and even financial levels of a population. Such an army represented a misuse of talent of the very worst kind. Some of the most intelligent men of their generation died on the Somme, caught on the wire and machine-gunned to death. Any one of them could have devised a better way of war than the generals who sent them to the slaughter. It was the nadir in personnel selection.

1

Fight or Flight

Fighting Spirit and Fanaticism in Medieval Warfare— The Battle of Nicopolis, 1396 and the Siege of Jerusalem, 1099

Medieval thinkers were quite clear about what they meant by courage. They differentiated between what they referred to as "true bravery" and "rashness." St. Thomas Aquinas believed that courage consisted of a firmness of mind in the accomplishment of duty. It was a virtue which made a man intrepid in the face of every danger, but there was an important proviso: it must be free of rashness. Courage was to the medieval mind midway between audacity and timidity. It was very much an aristocratic virtue, a noble form of behavior linked to race, blood, and lineage.

Yet courage was expected to rise above personal ambition and knights who were arrogant or rash—"outrageux"—were not considered courageous. To throw away one's life to no apparent purpose was also not necessarily a valiant act. Henry of Ghent gave an example of a Frankish knight who, at the Fall of Acre in 1291, when other knights were fleeing, threw himself into the midst of the Saracens and died fighting. The feeling was that his inspiration had been recklessness or a desire for personal glory and was not, therefore, a truly courageous act.

Pride and a contempt for the Turkish adversaries only slightly greater than that felt for their own Hungarian allies, caused the defeat of the French crusaders at Nicopolis in 1396. Arrogantly insisting that they would drive the Turks out of Europe, they boasted that "if the sky were to fall they would uphold it on the points of their lances." But the French knights were strangers to Eastern warfare and should have listened to their more experienced Hungarian allies. The fact that they did not indicates that they suffered from an ethnocentrism which led them to seriously undervalue the military prowess of others. Viewing the Turks as such insignificant opponents, they felt no need to take account of their enemy's strength or intentions. They disregarded the appeals by King Sigismund for caution and relied instead on their own undoubted bravery. But their overconfidence was also apparent in their lack of discipline. On the march to the city of Nicopolis the "frivolity" of the French knights and the pillaging of local towns and villages outraged their allies.

When King Sigismund held a council of war he told the French knights that he intended them to occupy the front line of the Crusader cavalry but to be preceded into battle by Wallachian foot soldiers, who would clear the field of the rabble of peasant conscripts that always fought at the front of a Turkish army. These foot soldiers, he explained, were used by the Turks to absorb the pressure from the charge of Christian knights so that when they tired they could be counterattacked by the lighter Turkish horsemen. This was good military sense, gleaned from a long exposure to the Turkish way of warfare. However, the French reaction was predictably unreasonable. They declared that they had not come so far and at such expense to go into battle behind a rabble of cowardly foot soldiers. The Constable d'Eu declared, "To take up the rear is to dishonor us and expose us to the contempt of all." As constable he was entitled to the front place in battle; it would be a mortal insult to have anyone ahead of him.

When the Turkish vanguard was sighted, Sigismund sent his grand marshal to plead with the French knights not to act rashly but to work in conjunction with him on a plan of battle. It seems that some of the French leaders like Coucy and Vienne were prepared to comply, but the constable shouted that the king wanted to deprive them of the glory and keep it all for himself. Admiral Vienne replied, "When truth and reason cannot be heard, then must presumption rule." With that the constable led the French contingent out to battle.

Although they were poor tacticians there was no denying that the French knights were formidable fighters. Properly used they would probably have won the day for Sigismund. As it was, the impact of their charge broke the first line of Turkish infantry but then came up against a line of sharpened stakes behind which the Turkish archers stood. Facing volleys of arrows, the French somehow broke through by sheer strength and hard fighting. Even then it was not too late for them to pause and wait for the main Hungarian army to come up and complete the victory. Coucy and Vienne again urged this, but by now there was little discipline in the French ranks and the constable refused to halt. Contemptuous of the Turks, the French had not bothered to ascertain the strength or composition of the Ottoman army. D'Eu believed they had broken their main force whereas, had he listened to Sigismund's advice, he would have realized that he had only beaten the vanguard. It was now that the foolhardy Crusaders found themselves surrounded by the full strength of the Turkish cavalry. What followed was a catastrophe in which the whole of the French contingent was either killed or captured. As Sigismund commented later, "We lost the day by the pride and vanity of these French; if they had believed my advice we had enough men to fight our enemies."

The foolishness of the French chivalry at the battle of Nicopolis in 1396 was no evidence of fighting spirit on the part of the French; more it was a sign of arrogance, stupidity, and even snobbery. They were to repeat their mistaken attitude at the battle of Agincourt against the English in 1415 and suffer a similar fate at the hands of English peasants and yeomen, falling to men who were their social inferiors but their military masters

There are occasions when too much personal courage is as harmful as too little, and where high morale can lead one into rash actions. When these military virtues are linked to religious fanaticism of the most extreme kind, one has a formula for disaster, which is exactly what happened to the Knights Templars in 1187 at the Springs of Cresson in Galilee.

Crusader contempt for the Muslims was at the root of many of the military reverses they suffered in the Holy Land. For sheer crassness it is difficult to find a better example than that of the master of the Templars, Gérard de Ridefort, at the Springs of Cresson. Charged with a diplomatic mission, Gérard was traveling toward Tiberias with a small group of companions. When he heard of the presence in the locality of a Saracen army of 7,000 mounted warriors led by Saladin's general, Keukburi, he immediately collected as many Christian soldiers as he could find at short notice. With just 140 knights and 300 foot soldiers he decided to attack Keukburi, who was blissfully unaware of Gérard's presence. In spite of the strongest arguments of Roger de Moulins, the master of the Hospitallers, that the disparity of numbers was too great, Gérard was not deterred. He asserted that the superiority of Christian knights was such that victory was certain. His own deputy, the marshal of the Templars, James de Mailly, sensibly recommended retreat. But Gérard accused him of being a coward and of loving his blond head too much to risk losing it. James retaliated by saying that he would die in battle a brave man, but that Gérard would flee like a traitor—a prediction that was to prove correct. Gérard was so impetuous that he charged the Saracens without even waiting for his infantry, who played no part in the fight. The Christian knights were first surrounded and then massacred, with James de Mailly and Roger de Moulins among the dead; but Gérard escaped with just two other Templars.

* * *

Crusaders of the eleventh and twelfth centuries were, by their very title, zealots. Their spiritual convictions inspired them to acts of great faith and awesome finality. Their beliefs allowed no room for doubt and the justice in their temporal actions rested in their spiritual certainty. When they lifted the sword in the defense of their religion it took on more than earthly power; its path became divinely ordained. Their successes were God-given; their failures were mere demonstrations of a lack of intensity in their faith. For the noble lords who took up the cross in 1096, it was a duty owing to their rank, that they must lead Christ's flock to free Jerusalem from the Muslims.

The journey to the Holy Land was a terrible test of faith. Many fell by the wayside, but still more struggled on secure in the knowledge that they did God's work. The greater the hardships they faced, the more certain they became in their purpose. No fighting spirit could be greater than that with which the soldiers of the First Crusade fought. Their belief in ultimate victory was certain; their own fears were light in the knowledge that anyone who fell in the struggle against the infidel would earn forgiveness of all sins. No Christian could die more worthily. Thus armed, the Christian soldier fought with absolute conviction. God was not only on his side, he was in his very arm as his sword rose and fell.

By 1099 the soldiers and pilgrims of the First Crusade reached Jerusalem. Their morale, which had slumped at several times on the long journey from France, had been boosted by miraculous revelations, like the discovery of the "Holy Lance" after they had captured the city of Antioch. Some of their victories against powerful Muslim armies had seemed miraculous as well, and by the time they reached Jerusalem they were so inspired that they would have endured any hardship to achieve their final goal. In the extracts that follow, from my book *Saladin and the Fall of Jerusalem,* I hope to demonstrate the inspiration provided to the Crusaders by their religious convictions. The fanaticism of the religious warriors—both Christian and Muslim—drove medieval soldiers to great acts of courage as well as terrible acts of cruelty.

> At noon, the hour Christians traditionally associate with the crucifixion of Jesus Christ, the (movable) siege tower of Godfrey of Bouillon was in position at the eastern end of the northern walls of Jerusalem. At this point, east of Herod's gate, the walls were some 50 feet high and the tower overlooked them by some seven to ten feet. In the upper section of the tower were Duke Godfrey himself, his brother Eustace of Boulogne and a company of knights; while in the middle section were Ludolph and Engelbert of Tournai. Amidst the noise of war—the crashing of great boulders on the timber of the tower, the crackling sounds of burning thatch, the curious whirring and hissing noises as the pots of "Greek fire" flew through the air like shooting stars, emitting fiery tails—the grim crusaders had little time for reflection. As they crouched low to the floor the siege tower was moved to within a few feet of the walls of Jerusalem. The goal of their journey and all their suffering was now just feet away; and yet these last few steps would be the hardest of any they had taken since leaving France two years before. Crouching, and occasionally crossing themselves or wiping the sweat from their eyes in the burning heat, the Frankish knights could sometimes make out the faces of the Muslims on the walls, faces contorted by fear and hatred, just as theirs must have seemed to their enemies. They were hardened soldiers, bred to their trade, veterans of a dozen such sieges and many battles in the field and yet this was very different. They had fought for their faith at Dorylaeum and Antioch but it had not been like this. The Muslims who manned these walls seemed in their eyes less than human, mere servants of the Antichrist who were fighting now to prevent true believers from inheriting the city of God. Here, in Jerusalem, there was to be salvation for all, forgiveness for sins, cures for physical and mental ills, and in the Church of the Holy Sepulchre a tangible link with their saviour Jesus Christ. In the hearts of the crusaders there was a hatred more bitter than any they had known before.
>
> The Muslim warriors on the walls—professional soldiers of the Fatimid caliph—knew none of this. They were fighting to defend their city from these Frankish barbarians. It

> is doubtful if they gave much thought to why the Christians had come and if they did it was simply to dismiss such reasons as the misguided superstitions of infidels. However, within the city, among the ordinary people, there was a thrill of fear, for news of the prodigious strength and size of these Frankish invaders had already reached them from cities that, like Antioch, had succumbed. Once their walls were breached they could expect little mercy from the followers of the Christian God.[20]

The motivation of the Crusaders was unlike that of warriors at any other time in history. They were not interested in loot, as was usual in those days, and willingly destroyed the possessions of those who defiled Jerusalem's holy places. Significantly, after they entered the city, the Christians ripped all the luxurious silken hangings of the Muslims, tipped their fine wines into the gutter, and killed their animals, even fine Arabian stallions.

> The Muslim defenders made a final effort to halt the inexorable progress of Godfrey's tower. Ropes were thrown over the wooden leviathan and efforts made to topple the whole structure, while rocks and pots of naphtha crashed into its walls of hide. But the crusaders managed to keep the tower intact by scything through the ropes which threatened it, and by using vinegar to extinguish the flames of a combustible log swung out against the tower by the defenders. In the smoke and confusion opposite them the Franks began to detect a slackening resistance. At once, Ludolph and Engelbert pushed out tree trunks from the middle section of the tower to the top of the walls and clambered onto the battlements, soon followed by Duke Godfrey and his knights. With a shout of triumph from the waiting troops below, a dozen ladders were placed against the walls and selected soldiers now scaled them to join the Lorrainers on the battlements. It was hand-to-hand fighting now in which there was little skill. Men wrestled with each other, sometimes toppling together from the battlements, sometimes gouging at each other's eyes or tearing at throats. The press was too close for sword strokes and handles were used like

[20]G. Regan, *Saladin and the Fall of Jerusalem,* pp. 7–8.

> hammers. But the longer Duke Godfrey and his men held the wall, the longer it gave the Lorrainers and Normans to use their ladders, while across the perilous bridge of logs first tens, then hundreds, of warriors flooded onto the walls. Soon, by sheer weight of numbers, the Muslims were pressed back and they turned to flee. At last Godfrey of Bouillon raised his banner above the walls as a signal that the city had been entered.[21]

The Muslim defenders of Jerusalem were professional soldiers from Egypt and not zealots like the Christians. They were fighting for pay rather than their religion and had none of the inspiration that their enemies took from their faith. They fought bravely but only in the way they would have fought their fellow Muslims had the need arose. The Christians fought under banners provided by Pope Urban II, and the bishop of Le Puy frequently produced holy relics to inspire the simpleminded soldiers. The territorial ambitions of the Crusade's leaders, like Godfrey of Bouillon, were kept entirely secret from the masses, who felt that all were equal in the sight of God.

> While Godfrey and his troops secured the northern walls and fought their way deep into the Jewish quarter of the city, Robert of Normandy and Tancred headed for the Temple Mount. Here the Muslim defenders held out for a few hours in the al-Aqsa Mosque. Although Tancred had promised them their lives and left his banner with them as a token, the Christian soldiers forced aside his guards and massacred everyone so that "they were wading in blood up to their ankles." The soldiers were inspired by such a fanatical hatred of both Muslims and Jews that as soon as the individual Christian leaders lost control of them in the narrow streets of the city a dreadful massacre developed which ranks with the most notorious in history and forever coloured relations between Christianity and Islam. It is easy to stand in judgment on the simple soldiers who followed the cross to the Holy Land and fought their way through incredible hardship to the walls of the Holy City itself. Mod-

[21]G. Regan, *Saladin and the Fall of Jerusalem,* pp. 8–9.

> ern examples of genocide should warn us of the power of religion and ideology to indoctrinate whole peoples. The men who slaughtered the Muslim population of Jerusalem were inspired by apparently higher motives than their leaders. Where the dukes and the barons looked for a chance to take prisoners and earn ransoms as was customary in European warfare, the common soldiers thought only that in this city their Saviour, Jesus Christ, had been crucified by the Jews and that its holy places had been desecrated by the Muslims. They had come not only to liberate Jerusalem but to cleanse it and this could only be achieved by a general slaughter.[22]

The cruelty of the Christian masses was a result of their religious fanaticism. While their masters might have been willing to grant ransoms to the more wealthy Muslims, the common soldiers had passed out of their control. Fueled by a religious fervor of almost-unimaginable intensity, the Christian soldiers slaughtered the defenders as defilers of God's holy city, showing them no mercy. The terrifying faith inspired in them by Pope Urban II at Clermont in 1095 was a force that could not be controlled, but could only be channeled into acts of barbarism like the slaughter of the Muslim population of Jerusalem. The Crusaders, half-starved and almost mad with the frustration of being denied immediate access to the holy city, fought with unparalleled ferocity. Individually they might have seemed weak, but as a group they gained strength from a complete unity of purpose. Their morale was like that known to no other army.

> Thus Jerusalem had taken on a more than earthly significance. To the simple men of the eleventh century the very name conjured up visions of the heavenly city itself, with gates of pearl and precious stones, as they had heard described by their village priests in France. Jerusalem was the centre of the spiritual world just as medieval maps depicted it as the centre of the physical world. In here the people of the world, scattered for so long, met again in pilgrimage.

[22]G. Regan, *Saladin and the Fall of Jerusalem,* pp. 9–10.

It was the place to which the elect of God ascended, the resting place of the righteous, the city of paradise. To the uneducated minds of the landless poor (who had followed the knights in 1099) Jerusalem was the city of eternal bliss.[23]

What followed the siege was a holocaust:

For the rest of the day and through the night the killing went on as the crusaders hunted down every living thing: man, woman, child, even animal. Much of the slaughter was carried out by the pilgrims who had accompanied the Crusade and now fell to their work with any weapon that came to hand: axes, clubs and even sharpened staves. Almost intoxicated by the killing, they hacked at everything in their path. By torchlight the Muslims of Jerusalem were hunted down, some dying by fire, some by the sword and others, abandoning hope, chose to leap to their deaths from the highest buildings. To the disgust of the knights, the Christian Tafurs—followers of the Norman knight known as "King Tafur"—even slaughtered the beautiful Arab stallions which were kept in the city. It was as if they could not rest until everything that had profaned the holy places had been slain. These terrible fanatics, who fought naked with faith alone as their shield and ate the flesh of their victims, drove the Muslims before them in greater terror than even the armoured knights of Duke Godfrey or Raymond of St. Gilles. As they moved through the city it was as if the Angel of Death had passed by leaving nothing living. Even the pots and jars of oil or grain were smashed, and sacks of corn ripped open like the bellies of their human victims.

In the Muslim holy places of the al-Aqsa Mosque and the Dome of the Rock, the bodies lay so thickly that they formed a veritable mound of flesh. Moreover the Christians had hacked their victims horribly, slashing open their stomachs in search of the gold coins that, it was rumoured, the Muslims had swallowed to avoid losing them to the invaders. The result was that the blood was literally ankle deep in some areas of the city, where drainage was impossible. For the Jews a different fate was reserved. They were herded together into their chief synagogue and burned alive in the

[23]G. Regan, *Saladin and the Fall of Jerusalem,* pp. 10–11.

great conflagration that followed the assault on that building. It was as blood-soaked butchers that the crusaders eventually went to the Church of the Holy Sepulchre to celebrate their victory and to thank God for his great goodness, while around them in the streets were the unburied corpses of the Muslim population of Jerusalem, the first martyrs in the jihád that was to rage for the next two centuries. William of Tyre records the impact on the crusaders themselves.

> It was impossible to look upon the vast numbers of slain without horror; everywhere lay fragments of human bodies, and the very ground was covered with the blood of the slain. It was not alone the spectacle of the headless bodies and mutilated limbs strewn in all directions that roused horror in all who looked upon them. Still more dreadful it was to gaze upon the victors themselves, dripping with blood from head to foot, an ominous sight which brought terror to all who met them.[24]

Even as cultivated and civilized a churchman as William of Tyre could justify the actions of the Christians as the just judgment of God, because these same Muslims had profaned his holy places with their rituals. The bloodshed and the slaughter, however shocking, expiated this sin. Although they would have found it difficult to express their views so clearly, this was how the ordinary soldiers had felt while they were killing. They were the tools of God, carrying out his purpose, and were guiltless of any responsibility for what they had done.

[24]G. Regan, *Saladin and the Fall of Jerusalem,* p. 11.

2

The "Bloodybacks" at the Battle of Minden, 1759

The British redcoated infantryman of the eighteenth century was a remarkable soldier. In media terms he has had a bad press. In Scotland he is seen as a persecutor of the highland clans—in Hollywood terms the enemy of Rob Roy McGregor. In the United States he is the brutal oppressor of American liberties and the enemy in the War of Independence. With his overfussy eighteenth century uniform, his tricorn hat, leather stock, and powdered wig, he is sometimes seen as effete, or at least hardly a match for the American woodsman or Indian warrior. The truth, of course, is very different. The British redcoat was very much a creature of eighteenth century European warfare, modeled on the Prussian soldiers of King Frederick William I of Prussia and his brilliant son, Frederick the Great. He was drilled to become part of a unit of concentrated fire-power, not an individual. The massed armies of European wars required little initiative on the part of their soldiers. The more a man thought, the more he fell prey to his fears. No thinking man would willingly stand upright no more than thirty paces from the enemy exchanging musket fire until he fell, or stand ready to receive bayonet thrusts from the enemy infantry or saber slashes from the passing cavalry. A thinking man would look for cover, or for a way of escape. And once a soldier thought of how to preserve his own life his efficiency became suspect.

During the middle decades of the eighteenth century, the British redcoat—derisively called "bloodybacks" by the

Americans on account of the frequency that flogging was used in the British army of the time—showed the two sides of his nature on battlefields in both hemispheres. At one moment running in panic from the Highlanders with their claymores, he next routed the cream of the French infantry at Fontenoy. Giving way to his fears on the Monongahela River in North America, the redcoat achieved at Minden in 1759 one of the most heroic and extraordinary victories in recorded history. The reason for this unpredictability, perhaps, rested with the quality of leadership the men received. Where the conditions of the fight were unusual it was most important for the officer to reassure his men and help them to fight to their strengths. Concentrated firepower was clearly a battle-winning weapon, but the stupidity of Braddock's officers in stressing this factor when the enemy was as elusive as forest Indians was fatal.

At Fontenoy in 1745, a British defeat incidentally due to the incompetence of the commander, the duke of Cumberland, the British infantry covered itself in glory. Frederick the Great of Prussia once acknowledged that in the confusion of battle the greatest contribution a commander could make was to make certain that he placed his men in a position that allowed them to make the most effective use of their first few musket volleys. Once casualties were suffered the effectiveness of the volley would fall off. It was therefore crucial that the first volley should be as powerful as possible. It might very well determine the outcome of the entire battle. If awards were given in some Prussian school of excellence for the best volley fired in an eighteenth century battle, then almost certainly the first prize would go to Lord Charles Hay's 1st Guards regiment at Fontenoy. Although the story is undoubtedly apocryphal whereby the British and French officers, displaying old world courtesy, insisted that the other should fire first, there was something very slow and ritualistic about the first infantry volley in battles at this time. It was almost equivalent to killing in cold blood, with regiments situated no more than thirty or forty paces apart and simply standing upright to receive fire. It was inhuman, which accounts for the dehumanizing drill that was necessary to force men to stand and watch the bullets approach them with their names written on them.

At Fontenoy, Lord Charles Hay, apparently bowed to the French officers and took out a hip flask and slowly drank from it. He then called out to the French, goading them: "I hope, gentlemen, that you are going to wait for us today, and not now swim the Scheldt as you swam the Main at Dettingen." It is unlikely that he asked them to fire first, for that would have been madness. However, the French did fire first, and very badly, with most of the bullets flying over the heads of the British front line. Whilst they reloaded they were virtually helpless. At Hay's order a rolling thunder of fire broke out, which everyone present agreed marked the most perfect volleys of musket fire anyone had ever heard or seen. The result was astonishing. The front line of the French infantry collapsed like a line of marionettes with their strings cut. Six hundred men of the French and Swiss Guards went down. In the words of Sir John Fortescue:

> The British infantry were perfectly in hand; their officers could be seen coolly tapping the muskets of the men with their canes so that every discharge might be low and deadly; while the battalion guns also poured in round after round of grape with terrible effect. The first French line was utterly shattered and broken. Even while the British were advancing, [Marshal] Saxe had brought up additional troops to meet them and had posted regiments Couronne and Soissonois in rear of the King's regiment, and the Brigade Royale in rear of the French Guards; but all alike went down before the irresistible volleys. The redcoats continued their triumphant advance for full three-hundred yards into the heart of the French camp.[25]

During the Jacobite Rising of 1745, the British redcoats—under a very different quality of leadership—showed another side of themselves. At the battle of Prestonpans on 21 September, the Royal commander Sir John Cope had

---

[25]J. W. Fortescue, *A History of the British Army, Volume 2,* p. 146.

the most extraordinary difficulty in making them fight at all. The men who could outface the French were, apparently, terrified at the savagery of the Highlanders they had to face. After firing just one volley—a desultory one as we learn—many of the redcoats threw down their guns and ran. Cope berated them: "For shame, Gentlemen, behave like Britons. Give them another Fire and you'll make them run. Don't let us be beaten by such a Set of Banditti." But nobody was listening. The redcoats were frightened of the Highland claymore, which was their nemesis. After the battle it was said that the field "presented a spectacle of horror, being covered with heads, legs and arms, and mutilated bodies . . ." Morale in the royal army slumped through simple fear. It was not just death the redcoats feared, but it was the kind of death. More than a century later British redcoats suffered the same sort of terror in fighting the Zulus in 1879, because the fierce tribesmen split the stomachs of their victims with their stabbing-spears.

After the fiasco at Prestonpans a public inquiry was held and Cope was questioned as to why his men had broken and run. Cope had no answer, except that they had been seized by "a sudden Panick." It was probably the speed with which the Highlanders attacked in contrast with the slow, methodical approach of Continental armies. Throwing discipline to the winds, the Scots simply tore into their opponents and hacked them to death. Only a truly disciplined force could withstand them, as the duke of Cumberland's did at Culloden in 1746, after which the Highlanders could be slaughtered at leisure.

The duke of Wellington wrote in later years of the British soldiers as the "scum of the Earth," and General Wolfe was no better impressed with his own men in 1758. As he wrote, "The condition of the troops at Portsmouth excels belief. There is not the least shadow of discipline, care or attention . . . Dirty, drunken, insolent rascals, every kind of corruption and immorality, and looseness carried to excess; it is a sink of the lowest and most abominable vices." Nevertheless, these were the same men who climbed the heights of Abraham in the dark and put the French army of Montcalm to flight once they got there, thereby securing Quebec—and, in point of fact, Canada—for the British Empire. Wolfe found his officers no more prepossessing, calling them "an effeminate race of coxcombs." Nevertheless, these "chinless wonders"—of whom, incidentally, Wolfe was one—seemed to be able to inspire their "drunken, insolent rascals" in the most extraordinary way. For generations, popinjay officers led British armies into battle, revealing the most extraordinary courage, if usually lacking the professional military skills of their enemies. How often during the First World War did young men of just eighteen years, straight out of school and with hardly any officer training, lead their soldiers into the face of German machine gun fire armed only with umbrellas, walking sticks, or cricket bats? The Germans viewed such antics as the height of folly, but then courage in battle is perhaps the supreme kind of folly anyway. And the heroism of the redcoat at Fontenoy and Minden became part of Britain's military tradition. Steadiness with musket and bayonet became the hallmark of the British soldier and at home the public came to admire not the flashy antics of continental hussars and cuirassiers in their gaudy costumes, but the imperturbable, phlegmatic, enduring discipline of the redcoat, and his unshakable courage. In this way, the British liked to believe that the redcoats were a mirror of the nation.

There are certain parallels between Sir John Cope's defeat at Prestonpans and Braddock's disaster on the Monongahela River in 1755. In both cases, the commanders underestimated their enemies, judging both Highlanders and Indians by the European standards of the day. In view of the fact that their opponents were irregular fighters of extreme savagery, both Cope and Braddock should have approached matters with greater caution. In both cases the redcoats were not the best troops of their type, both forces being prey to their own fears and magnifying the accomplishments of their enemies. The "unknown and the unexpected" easily cracked their fragile discipline, and the redcoats had nothing to fall back on, even personal initiative or *esprit de corps.* In Braddock's case the British regiments he had with him—the 44th and 48th—had lost some of their regimental spirit by dilution; they had been brought up to strength by reinforcements from other regiments—probably the men nobody else wanted—and by recruiting riffraff from among the American colonists. It was impossible to restore equilibrium to a regiment that has been treated in this way. Moreover, many of the redcoats were simple Irish lads, who found the gloomy American forest a depressing place. Their fears were fed by stories of Indians which made them seem more like fiends or monsters from children's stories. The soldiers "painted a picture" with themselves as victims. When the crisis came, and the Indians ambushed the British column, Braddock and his officers insisted on using European tactics in a North American environment. While the Indians hid behind trees, the British lined up to fire volleys. The Indians sniped their victims, the British peppered the undergrowth with bullet and shell. Those redcoats who tried to take cover were beaten back to their formal lines by their officers and so were shot down by an "invisible foe." The failure of the redcoat on this occasion was a failure of preconception: their commander had preconceived notions on the superiority of formal discipline against informal enemies, and the soldiers themselves were fighting not human beings but the creatures their fears had fashioned. When their discipline cracked it was because beneath the drillmaster's veneer there was nothing but a terrified and inadequate Irish farmboy, baffled by a situation he could never have imagined.

The battle of Minden, fought on 1 August, 1759, was for Britain one of the most significant battles of the eighteenth century. Politically, it was a battle fought to save Hanover being overrun by the French, but militarily it contained two incidents representing the extremes of the military spectrum. It provided one of the most famous acts of cowardice when the commander of the British cavalry, Lord George Sackville, refused to obey a direct order to attack from the commander-in-chief, the duke of Brunswick, for which he was later court-martialed and driven out of the service. The battle also contained an episode of almost-suicidal bravery by the British infantry who, though outnumbered many times over, attacked the massed French cavalry and drove it off the field. So astonishing—even unprecedented—was this action that the French commander, the Marquis Louis de Contades, remarked, "I have seen what I never thought to be possible—a single line of infantry break through three lines of cavalry ranked in order of battle and tumble them to ruin."

Confining our attention to the struggle between the British redcoats and the French cavalry, it is necessary simply to observe that in total the French army numbered some 60,000 men with 162 guns, while the Anglo-German army of the duke of Brunswick fielded just 45,000 with 170 guns. The right of Brunswick's army comprised six British infantry regiments—the 12th (Napier's), 20th (Kingsley's), 23rd (Hulke's), 25th (Howe's), 37th (Stuart's) and the 51st (Brudenell's)—as well as a Hanoverian regiment, all under the command of Lieutenant-General von Spoercken. It was formed in two lines; the first—under Waldegrave—consisted of the 12th, 37th, and 23rd, and the second—under Kingsley—of the 20th, 51st, and 25th. Each of the British regiments had had its best soldiers—the grenadier company—removed for use on another part of the field. As a result, the regiments averaged only about 500 men each at the start of the action.

Like the infamous "Charge of the Light Brigade" a century later at Balaclava, the drama at Minden began with a misunderstood order. But rarely could a mix-up have had a better or more extraordinary outcome. At just after dawn on 1 August, one of the duke of Brunswick's aides arrived at General von Spoercken's headquarters with the order that he was to advance at the "sound of the drum." There was nothing unusual in this, for coordinating troops on an eighteenth century battlefield was frequently carried out by drumbeat. All Brunswick was saying was that when he wanted the British infantry to advance he would signal them by a drum call. Incredibly, when von Spoercken passed the message to the British general Waldegrave, speaking in French as was the custom at the time, Waldegrave mistranslated what his commander had said and thought he meant that the British must advance "to the sound of the drum," meaning that they would advance immediately with drums beating. At once Waldegrave ordered his drummers to sound the advance and the first British line set off toward the French. The second line was amazed to find their colleagues marching away and promptly set off after them. Observing from far off the splendid sight of the perfectly aligned troops in their white gaiters, red jackets, and black tricorn hats, Brunswick might almost have been excused for saying *"C'est magnifique mais ce n'est pas la guerre."* Instead he raged and fumed. He could see his chances of victory being thrown away. Any moment the French artillery would tear the British lines to shreds. And for a while this is exactly what seemed to be happening. While the two lines of infantry headed straight toward the French cavalry in the center of the French army, the French gunners, backed up by regiments of infantry, poured a deadly fire into them. Several of the survivors left behind vivid accounts of what happened. Lieutenant Thompson wrote:

> On the immediate sight of us, they opened a battery of eighteen heavy guns which from the nature of the ground, which was a plain, flanked this regiment in particular every foot we marked. Their cannon was ill-served at first, but they soon felt us and their shot took place so fast that every officer imagined the battalion would be taken off [destroyed] before we could get up to give a fire, notwithstanding we were then within a quarter of a mile of their left wing. I saw heads, legs and arms taken off. My right-hand file of men, not more than a foot from me, were all by one ball dashed to pieces and their blood flying over me, this I must confess staggered me not a little but, on receiving a confusion in the bend of my right arm by a spent musket shot, it steadied me immediately, all apprehensions of hurt vanished, revenge and the care of my company I commanded took [their] place and I was *then* much more at ease than at *this* time. All the time their left wing was pelting us with small arms, cannon and grape shot, and we were not suffered to fire, but stood tamely looking on whilst they at their leisure picked us off as you would small birds on a barn door. I cannot compare it with anything as their shot

came full and thick, and had one quarter of them taken place [i.e., hit their target] there could not have been a man left.[26]

Another officer, Lieutenant Hugh Montgomery, remembered the same terrible advance into the heart of the French fire:

> Now began the most disagreeable march that I ever had in my life, for we advanced more than a quarter of a mile through a most furious fire from a most infernal battery of 18-pounders, which was first upon our front but, as we proceeded, bore upon our flank, and at last upon our rear . . . At the beginning of the action I was almost knocked off my legs by my three right-hand men, who were killed and drove against me by a cannon ball—the same ball also killed two men close to ward, whose post was in the rear of my platoon . . . It might be imagined that this cannonade would render the regiments incapable of bearing the shock of unhurt troops drawn up long before on ground of their own choosing, but firmness and resolution will surmount almost any difficulty.[27]

Rarely can firmness and resolution have been in greater demand. Yet, the fighting spirit of the soldiers seemed to increase at every step. As dozen after dozen of redcoats were literally torn to pieces by the gunfire, the men never wavered or hurried, yet within them was the urge to strike back at their tormenters. As one of them wrote, "The soldiers, so far from being daunted by their falling comrades, breathed nothing but revenge. Though at the beginning of the engagement I felt a kind of trepidation, yet I was so animated by the brave example of all around me that when I received a slight wound by a musket ball slanting on my left side, it served only to exasperate me the more, and had I then received orders, I could with the greatest pleasure have rushed into the thick of the enemy."

All this time the French cavalry commander, General

[26]E. Sanger, *Englishmen At War,* p. 167.
[27]Ibid., p. 168.

Fitzjames, had been watching the two lines of infantry approach, at any moment expecting them to break under the deluge of shot and shell that had been hitting them. But, puzzlingly, they continued to advance toward him without even attempting to form squares, the normal defense when faced with cavalry. Lines of infantry had never before been known to stand against cavalry. Fitzjames concluded that the British officers must be mad or incompetent, or both. He had sixty-three squadrons of cavalry drawn up in three lines and he ordered the first lines—eleven squadrons under the marquis de Castrie—to put an end to the impudent Englishmen. To the sound of trumpets the French cavalry rode down on the two red lines ahead of them. At the head of the first, Waldegrave calmly ordered his men to prepare to fire. Not until the horsemen were within ten paces did the order to fire ring out. Fire discipline, under the trying circumstances, could well have been ragged. But on this occasion the men fired as *one* and the French cavalry was shattered as if it had ridden into a brick wall. Hundreds of men and horses were downed in a split second and soon the front of the British line was a sea of writhing horses and screaming men. Even the British officers were horrified at the effect of such a volley at such short range. Thompson wrote, "such a terrible fire that not even lions could have come on, such a number of them fell, both horses and men, that it made it difficult for those not touched to retire."

In the French ranks a cry of both anger and amazement went up as the remnants of the cavalry veered away from the enemy; it was Crécy and Agincourt rolled into one. Now the second line of cavalry—an even bigger force—rode out to replace their worsted comrades. Brunswick saw that the only chance of saving the infantry—and perhaps even saving the battle—was for Lord Sackville to lead out the cavalry while the French were disordered. But now occurred the deplorable incident that made Minden a grim memory for the British cavalry just as it was so splendid a memory for the infantry. While Sackville argued with Brunswick's aides and found reason after reason for not obeying orders, the redcoats prepared to meet fourteen squadrons of fresh French horsemen. But the French had already made things difficult for themselves. The ground in front of the British lines was made difficult by the bodies of wounded and dead men and horses. Nevertheless, pride and shame drove the Frenchmen forward, but again the redcoats timed their volley to perfection and routed the horsemen at point-blank range. By now almost half of the entire French cavalry had been defeated by the first line of just three British regiments. The spirit of Waldegrave's soldiers was sky-high. They believed that they could do anything—perhaps even beat the whole French army on their own. However, Contades had decided to put a stop to the whole charade. While fifteen hundred elite troopers under the marquis de Poyanne charged the front of the British line, he ordered his infantry to attack each flank at the same time. The British would be overwhelmed by sheer weight of numbers.

If Sackville and the British cavalry were prepared to watch their comrades overwhelmed, the same could not be said of the British artillery, who chose this moment to enter the fray. Captain Macbean, with ten heavy guns, sized up the situation at a glance. As he put it, "They were going to gallop down, sword in hand, among our poor mangled regiments, but we clapt our matches to the ten guns and gave them such a salute as they little expected: for we mowed them down like standing corn." Poyanne's gendarmes had veered toward the Hanoverians, who had advanced alongside the British. But here they received as rough treatment as from the British regiments and soon hundreds of them were unhorsed. The French cavalry had been beaten more completely than on any field in anyone's memory and by infantry, who simply refused to be beaten. Yet even now von Spoercken's invincible infantry were not finished. Having dealt the cavalry a mortal blow, they now faced the advancing French infantry. First the 12th and 37th regiments routed the régiments de Condé, Aquitaine, and du Roi. As Montgomery explained succinctly: "We killed them a good many and the rest ran away." Next a regiment of Saxon grenadiers—Saxony was in alliance with France at this time—fell on the depleted ranks of the 12th and 37th. Montgomery described what followed: "As fine and terrible-looking fellows as I ever saw. They stood us a tug, notwithstanding they beat us off a distance, where they galled us much, they having rifled barrels, and our muskets would not reach them. To remedy this we advanced; they took the hint and ran away." Finally when more Saxons attacked, Spoercken decided it was time to use his second line, which was commanded by General Kingsley. Kingsley's three regiments had suffered the frustration of not being able to do much more than suffer casualties without striking back at their assailants. They soon put that to rights by shattering the remaining Saxon infantry and driving them "pellmell" from the field.

Robbed of what might have been one of the most complete victories in history by Lord Sackville's refusal to commit the cavalry to the contest, Brunswick had to be content with the partial victory that the British infantry had given him. Contades withdrew his army having suffered seven thousand casualties in dead and wounded, as well as five thousand prisoners. His cavalry had been so savaged by von Spoercken's redcoats that the French foreign minister, the duc de Choiseul later wrote, "The thought of Minden makes me blush for the French army."

3

"That Astonishing Infantry" at the Battle of Albuera, 1811

The duke of Wellington's description of his own troops is rightly famous: "the scum of the Earth." Yet he knew that for all their character faults—and there were many, one has only to think of the aftermath of the siege of Badajoz to realize that—there were no troops like them. On another occasion he said of them, "It all depends on that article [pointing at a passing British soldier] whether we do the business or not. Give me enough of it and I am sure." The French certainly knew what they were up against: in the aftermath of the battle of Albuera, Marshal Soult wrote to Napoleon, "There is no beating these troops. They were completely beaten, the day was mine, and they did not know it and would not run away." It was perhaps Napoleon's good fortune that he never met these troops in battle. Soult had told him that *"l'infanterie Anglaise en duel, c'est le diable."* Yet the British infantry at Waterloo was only a pale reflection of the men with whom Wellington had won battle after battle in Spain. This may account for Napoleon's rash and scornful description of the British at Waterloo, that Wellington was "a bad general" and the British were "bad troops." He said that the battle against the English would be *"un affaire d'un déjeuner."* If he believed that, he was in for a shock.

The men who fought at Albuera were far from being the "scum of the Earth." Many of them were quite sensitive to the horrors of war and have provided us with a clear appreciation of the mind of the soldier before he goes into battle. Moyle Sherer described the onset of fear: "My bosom beat very, very quick; it was possible, that the few minutes of my existence were already numbered. Such a thought, however, though it will, it must arise, in the first awful moment of expectation, to the mind of him who has never been engaged, is not either dangerous or despicable, and will rather strengthen than stagger the resolution of a manly heart." George Gleig spoke of the "apprehension" a man feels before a battle and the "feeling that takes possession of him":

> In the first place, time appears to move upon leaden wings; every minute seems an hour, and every hour a day. Then there is a strange commingling of levity and seriousness within himself—a levity which prompts him to laugh he scarce knows why, and a seriousness which urges him from time to time to life up a mental prayer to the Throne of grace. On such occasions little or no conversation passes. The privates generally lean upon their firelocks, the officers upon their swords; and few words, except monosyllables, at least in answer to questions put, are wasted. On these occasions too, the faces of the bravest often change colour, and the limbs of the most resolute tremble, not with fear, but with anxiety.[28]

The British soldier—villain and saint as he could be—showed the best of himself at the terrible battle of Albuera. With his commander—Marshal Beresford—frankly out of his depth, his allies uncertain, and his enemy overwhelming in strength and confidence, he held his ground and fought the French to a standstill before, with a final charge, he sent them rushing in precipitate rout. Albuera is rightly famous in the regimental histories of many British units, and has served to give inspiration to generations of young soldiers who must have wondered whether they could ever match the efforts of "that astonishing infantry."

On 12 May, 1811, General Beresford, marshal of Portugal, was besieging the Spanish city of Badajoz, heavily forti-

[28]P. J. Haythornthwaite, *The Armies of Wellington,* p. 204.

fied by the French, when he received news that a French army of 25,000 men under Marshal Soult was advancing to its relief. Abandoning the siege, Beresford pulled away to the southeast to the small town of Albuera, where he took up a defensive position. Three days later he received 14,000 Spanish reinforcements under the command of the Spanish generals Blake and Castanos, who agreed to serve under Beresford's command. The British general now had an army of about 37,000, though many of these were Portuguese, and the newly arrived Spanish troops were of questionable quality. Expecting the French to come from the east, Beresford made Albuera the center of his position. Unaware that Beresford had received reinforcements, Soult attacked Beresford's position on 16 May at dawn. The French made a feint attack on the center of the British line in the hope that Beresford would reinforce it, which in fact he did, by transferring troops from his reserve.

Meanwhile, the Spanish troops to the south of Albuera had just witnessed one of the most astonishing sights of the whole Napoleonic period—and certainly the strongest attacking force of the Peninsular War—in the shape of a massive column of 14,000 infantry, supported by artillery and 3,500 cavalry intent on turning the left flank of the allied position. The French column was 400 yards across and 600 yards deep. Its momentum should have been irresistible. The Spanish refused to face such an onslaught and many of them fled, though others showed great gallantry in a losing cause. To support the hard-pressed Spaniards, Beresford next ordered the British 2nd Division to advance and form a second line behind them. Instead of obeying these sensible orders, the division's impetuous commander, General William Stewart, ignored them and attacked the French without a second thought. His leading brigade of 1,600 men—Colbourne's—attacked the left flank of the huge French column, mowing down hundreds of the French with their first volley. Temporarily Soult's column was checked. Then fate intervened. A sudden rainstorm saturated the battlefield, rendering muskets on both sides useless. Coinciding with this blow, the French cavalry—Second Hussars and Polish Second Vistula Lancers—now advanced and overran Colbourne's men, who had no protection against horsemen. Soult's Polish lancers had a field day, killing hundreds of helpless British foot soldiers with sword and lance. Within five minutes 1,300 out of 1,600 of Colbourne's men were lost, as well as five regimental colors. Some of the cavalry even tried to capture the British commander himself. But Beresford—a huge man—simply seized one Pole's lance, picking the rider out of his saddle and throwing him to the ground.

Now Hoghton's brigade appeared on the scene, followed by Abercromby's, and they formed a line against the French column. Seven British battalions—about 3,700 men, two-deep—faced two French divisions of about 7,800 men in column in one of the most extraordinary firefights at close range in all history. In places muskets were touching each other, and the combatants were rarely more than twenty yards apart. Volley after volley was fired, but the British "line" held, even though Hoghton himself was killed, along with the three battalion commanders. It was at this moment that Colonel Inglis of the 57th called on his men to "Die Hard!" The British losses were dreadful: the 29th lost 336 out of 476; the 57th lost 428 out of 616; and the 48th lost 280 out of 646. French losses were never computed, though they must have been at least as heavy.

At this moment, when the strongest lead was needed from the commanders, both Beresford and Soult lost their nerve. Soult had just realized that the British had been reinforced and that he was outnumbered; he therefore went onto the defensive. Beresford, baffled by the murderous impasse between the rival infantry had matters taken out of his hands by one of his subordinates. Fortunately, with his commander temporarily frozen with indecision, the deputy quartermaster general, Major Henry Hardinge, took the initiative, calling up Sir Lowry Cole, commander of the 4th Division, to break the deadlock and save the British infantry, who were heavily outnumbered and would soon have been overrun. Cole brought forward 4,000 fresh infantrymen "in line," with a protective square at each end to resist the French cavalry if they tried to turn his flanks. The 4th Division was welcomed by a torrent of French artillery fire as well as a charge by French dragoons trying to turn their flank. Next, three French columns, supported by artillery, tried to break Cole's line. Outnumbered by three to one as he was, Cole was still able to concentrate more firepower than the French, as the British line formation enabled 2,000 British troops to fire at one moment against just 360 of the French, namely those at the head of their columns. The Fusilier Brigade, consisting of the 1st and 2nd Battalions of the 7th Regiment and the 1st Battalion of the 23rd, stood their ground and fought the most bitter fight in all British military annals. Sixteen-year-old Ensign Thomas was heard to yell, "Only with my life!" as a swarm of French cavalry tried to seize the colors of the Buffs. Thomas was killed and the color standard taken but Lieutenant Matthew Latham leapt forward and wrestled back the color from the French cavalryman who had it. Latham was hacked down by the hussars, but in falling he ripped the regimental color from its staff and bundled it into his tunic before passing out. Latham recovered, though he had lost an arm and much of his jaw in the fracas. He had saved the color and his efforts are forever remembered in the form of a silver replica of the incident which adorns the regimental dining table—an inspiration to every young soldier who joins the regiment.

The historian Napier relates the stirring efforts of the British infantry in his memorable account:

> The Fusilier battalions, struck by an iron tempest, reeled and staggered like sinking ships; but suddenly and sternly recovering, they closed on their terrible enemies, and then was seen with what strength and majesty the British soldier fights. In vain did Soult with voice and gesture animate his Frenchmen; in vain did the hardiest veterans break from the crowded columns and sacrifice their lives to gain time for the mass to open up on such a fair field; in vain did the mass bear up, and fiercely striving, fire indiscriminately upon friend and foe, while horsemen hovering on the flank threatened the advancing line. Nothing could stop that astonishing infantry. No sudden burst of undisciplined valour, no nervous enthusiasm, weakened the stability of their order . . . In vain did the French reserves joining with the

> struggling multitude endeavour to sustain the fight; their efforts only increased the irremediable confusion, and the mighty mass giving way like a loosened cliff, went headlong down the ascent. The rain flowed after in streams discoloured with blood and fifteen hundred unwounded men, the remnants of 6,000 unconquerable British soldiers, stood triumphant on that fatal hill![29]

There were no orders given, except "Close in! Close up! Fire away!" Bit by bit the French were pushed back. At last the French broke and a rout began. But the victory had been dearly bought. Never before or since had British soldiers fought so bravely. In places their bodies lay three-deep. The 2nd Battalion of the 7th Regiment had just 85 men left standing at the end out of 568! Sergeant Cooper of the 2nd battalion recounts what happened:

> Men are knocked about like skittles; but not a step backwards is taken. Here our colonel, Sir William Myers, and all the field officers of the brigade fall killed or wounded, but no confusion ensued . . . We are close to the enemy's columns; they break and rush down the other side of the hill in the greatest mob-like confusion.[30]

Beresford was too baffled by his victory to do any more than occupy the battlefield. He was very gloomy and wrote mournfully to Wellington to tell him about the battle. Losses among his Spanish and Portuguese troops had been minimal, but his British infantry had lost 4,407 out of 8,800, losses that could not be easily made good. Wellington, shocked as he was, decided that the "victory" needed to be written up, otherwise the politicians and the public in Britain would be demanding his head. A few days later, when he visited the battlefield, he saw the many British dead, lying in their ranks. He also visited the wounded, telling them, "Men of the 29th, I am sorry to see so many of you here." One veteran sergeant replied, "If you had

[29]M. Glover, *That Astonishing Infantry,* pp. 51–52.
[30]C. Messenger, *For Love of Regiment,* p. 133.

commanded us, my lord, there would not be so many of us here."

Yet who were these men who fought with such unparalleled bravery that their enemy commanders were almost their greatest admirers and yet of whom their commander said, "I don't know what they do to the enemy but they certainly frighten me"? In general they were the "restless spirits" among the lower classes, men who could not settle to a job and who would have as readily turned to crime as to the armed services if matters had taken a different turn. Recruiting posters sometimes tell us what sort of person was being targeted. This particular drive was aimed at prospective dragoons: "All you who are kicking your heels behind a solitary desk with too little wages, and a pinch-gut Master—all you with too much wife, or who are perplexed with obstinate and unfeeling parents." Infantrymen would have been raised from a lower order of candidate. The duke of Wellington would never have believed that his soldiers enlisted from feelings of personal pride or honor, although there is evidence that this was true of some. Most, however, joined up to secure a regular wage and food, and a roof over their heads. Some others, even, took the king's shilling to escape the king's justice. What united them all, though, was a feeling of superiority over foreigners, which foreign service overseas gave them an opportunity to demonstrate. It was this chauvinism—based as it was on an almost-complete ignorance of the merits of other peoples and cultures—that contributed to British "fighting spirit" in battle. Doctor Samuel Johnson wrote an essay on "The English Common Soldiers," which was published in 1801, in which he explained the success of British arms in this way: "Our nation may boast, beyond any other people in the world, of a kind of epidemic bravery, diffused equally through all its ranks." Yet, according to Johnson, this courage did not grow from esteem of their leaders for "It does not often happen that he [the British soldier] thinks much better of his leader than of himself . . . The Englishman despises such motives of courage: he was born without a master; and looks not on any man, however dignified by lace or titles, as deriving from nature any claims to his respect, or inheriting any qualities superior to his own." Johnson also warned his reader not to complain overmuch about the behavior of the soldier in peacetime. As he pointed out, "They who complain in peace of the insolence of the populace, must remember, that their insolence in peace is bravery in war."

It was not easy for allied troops—Portuguese, Spanish, or German—to achieve an understanding with their British counterparts. As one allied soldier wrote:

> The English are admired as a free, an enlightened, and a brave people, but they cannot make themselves beloved; they are not content with being great, they must be thought so, and told so. They will not bend with good humour to the customs of other nations, nor will they condescend to soothe (flatter they never do) the harmless self-love of friendly foreigners. No: wherever they march or travel, they bear with them a haughty air of conscious superiority, and expect that their customs, habits and opinions should supersede, or at least suspend, those of all the countries through which they pass.[31]

In essence, however lowly his origin, the British soldier was a snob. His sense of "conscious superiority" sustained him in combat with his enemies. His courage was a pride in race even more than an *esprit de corps.* He regarded his military traditions as more significant than the contrivances of foreigners and his victories more momentous. For all that the "Iron Duke" chose to denounce the raw material that made the clay of his profession, he was also on record as admitting, "I consider nothing in this country so valuable as the life and health of the British soldier." And his Peninsular veterans would have agreed with that.

[31]P. Haythornthwaite, *The Armies of Wellington,* p. 21.

4

"Victory or Death" at the Alamo, 1836

The defense of the Alamo by Texan volunteers in 1836 has become one of the epic stories of American history. Popularized by Hollywood, it almost ranks with the saga of Custer's Last Stand, yet in other ways it is a far more worthy story of courage and self-sacrifice. At the Little Bighorn, the Seventh Cavalry were under orders to follow their commander, however unwise these orders might be. Moreover, their "last stand" revealed the courage of desperation. Every man with Custer knew that there was neither chance of escape nor of surrender. The Sioux and Cheyenne Indians would have tortured to death any man taken alive. The troopers fought because, at the last, every man will fight for his life because he has to. Yet the defenders of the Alamo were volunteers who fought for a cause; one close to the hearts of all Americans and one for which they had died in the War of Independence and would die again in the great civil war to come. Fighting for the cause of freedom seemed to purge Americans of their cruder natures and bring out the best in men who had hardly led blameless lives before. It inspired men of limited intellectual or spiritual ability to acts of sublimity, in which they gave their own lives so that others might live in freedom.

During the last days of Spanish rule in Mexico, the authorities had encouraged settlers to move into the province of Texas and many settlers from the United States, encouraged by the offers of free land, moved over the border into Texas. One of the provisos, acknowledged as much in the breach as the compliance, was that the new settlers would become Roman Catholics and Mexican citizens. The sort of men prepared to enter Texas under the lure of free land were generally not much concerned about matters of religion and probably took the view that one could as easily ignore one faith as another. By 1835 as many as 28,000 Americans had settled in Texas, reducing the native population to a small minority. Although Mexico had won its independence from Spain in 1821—mostly through the achievement of General Agustin de Iturbide, who had had himself declared emperor—most Texans saw little sign of improved government now that their land was administered by Mexicans. Two years later a Republican coup led by Antonio López de Santa Anna Perez de Lebron, a former Spanish officer, overthrew the emperor and created the new Mexican Republic, with Santa Anna as the first president. The Americans in Texas—who referred to themselves as Texians rather than Mexicans—were predictably hostile. They believed that they would get a very much worse deal from the Mexicans than from the corrupt and lazy old Spanish system, and were prepared to resist any attempt by Santa Anna to interfere with their freedom.

Matters came to a head in 1835 when General de Cos, Santa Anna's brother-in-law, with fourteen hundred men, established a military base at San Antonio in Texas. At once the Texans felt threatened. Just across the river from the fort was the Alamo, a former mission and barracks. De Cos seemed determined to provoke an incident which would enable him to take military action against the American settlers. At last he found a suitable bone of contention. He insisted that the inhabitants of Gonzales should return an ancient cannon lent to them many years before to scare off Indians. The inhabitants at once made up a banner saying, "Come and take it," and refused to return the cannon, saying it was Texan property not Mexican. De Cos sent troops to seize the gun but on 2 October, 1835, the locals ambushed the Mexican troops and forced them to retreat. This somewhat incongruous incident began the war of Texan independence.

The Texans began forming an army at Gonzales, though at first it numbered just five hundred men and was riven by dissent. Every man wanted to be a commander—it was a case of "too many chiefs and not enough Indians." News of the rising soon reached the United States and the government there did nothing to discourage the despatch of money, weapons, and volunteers to help the Texans in their struggle. Many American veterans from the War of 1812 against Great Britain seized the opportunity to find more action in the south, though others were deserters from the American army who hoped to find adventure without too much discipline.

Meanwhile, the settlers at Gonzales had decided to force the issue by attacking the Mexican troops at San Antonio. Command of the small Texan army was given to James Bowie, a rich land speculator, famous as a duelist and for the knife he carried. He had settled in Texas after making a good marriage to the daughter of the provincial governor, becoming a Mexican citizen in the process. Bowie felt a sincere and deep commitment to the cause of Texan independence, although recently his addiction to hard liquor had reduced his effectiveness, both as a businessman and as a soldier. On 27 October, Bowie's men routed the Mexicans two miles south of San Antonio, killing sixty-seven of them and wounding a hundred more. The Mexicans retired into San Antonio, while Bowie awaited reinforcements. For the Texans, it was a victory, but as yet it was far from decisive. Weeks passed with little progress being made and as winter set in some of Bowie's men began to desert. However, the newly arrived American volunteers pressed Bowie to take a decisive step. On 6 December, three hundred of the Texans assaulted San Antonio. The fighting lasted for four days until de Cos offered to surrender. But the war was only just beginning and many of the American volunteers assumed they had won a victory that absolved them of any further need to stay in Texas. As a result there was a general desertion from Bowie's ranks. Ultimately, only 104 men under Colonel Neill remained in San Antonio, where they began fortifying the Alamo.

If Texas was to achieve independence, it needed strong military leadership and it was fortunate in finding a strong leader in Sam Houston. Houston, a lawyer by profession, had experience of fighting the British in the War of 1812, as well as of having been a former governor of Tennessee and general of Militia. He had the ability and experience to supply the Texans with more than mere bravado. He set about raising a force of men who would be trained for battle against European-style troops, as the Mexicans were, and who would be well supplied with uniforms, food, ammunition, and all the rest of the military regalia of the time. But more than this, Houston set about finding men who could act as his lieutenants, men who could accept discipline and could inspire it in their men. This was harder than it seemed. The Texan settlers were not natural soldiers. They had come to Texas in the first place in the hope of escaping the creeping disease of "firm central government." They were freedom-loving mavericks who did not intend to exchange American discipline for Texan. They would accept Houston as their leader as long as he did not start issuing orders. The problem was that Houston was aware that General Santa Anna was raising an army to suppress the revolt in Texas. Unless he could turn the Texans into something more than a group of individuals, the cause of Texan independence would be lost.

Houston ordered Bowie, now appointed a colonel, to return to the Alamo to dismantle its defenses and round up its garrison. It was Houston's intention to abandon the fort, but when Bowie arrived there he found the garrison on the point of mutiny. He soon restored morale by buying up provisions from his business contacts and, after discussions with Neill, decided not to abandon the Alamo. After all, Santa Anna would have to pass by with his army, and Bowie felt that he would have to spend time in capturing the Alamo before marching on, thereby giving Houston more time to build up his army. It was a brave decision.

When the news spread that Bowie intended to hold the Alamo against anything the Mexicans could throw at it, the imagination of Americans across the border was much provoked. As a result, reinforcements began to arrive at San Antonio. Lieutenant William B. Travis arrived first with thirty men, followed five days later by Colonel David Crockett and his Tennessee volunteers. Crockett was by then a famous frontiersman, Indian-fighter and tub-thumping politician. He was already a legend and his presence at the Alamo lifted the whole affair several notches toward the realm of fantasy and away from the dreary world of Texan frontier politics. And when Crockett declared that he had come to fight for "a cause," which was nothing less than the defense of "the liberties of our common country," the issue of Texan independence became an American, not a Mexican, issue.

While Neill was away from the Alamo, command was passed to Travis, as senior regular officer. Unfortunately, with so many "prima donnas" on hand to resent each other, Bowie objected to the command being given to Travis. In frustration he began to drink too heavily, destroying much of his health in the bargain. Travis, in an attempt to save both the "cause" and Bowie's health eventually agreed to share command with Bowie. But the Alamo was still short of men to garrison it and so Travis asked for assistance from Colonel Fannin at Goliad, ninety miles away.

Meanwhile, Santa Anna had raised a force of one cavalry and two infantry brigades, as well as engineers and artillery. By the standards of the Western hemisphere at that time it was a formidable army, made up of many regular soldiers. On 1 February, 1836, it left Saltillo and marched north. Morale among his men was surprisingly good and there was an air of confidence in his army. They felt that they were not so much going to suppress rebel Mexicans as to wipe out "Gringo invaders." After a three-week march they reached San Antonio, by which time Travis had moved all his troops inside the Alamo. Santa Anna was sufficiently impressed by the garrison's firepower that he abandoned the idea of an immediate assault and instead settled in for a siege. Inside the Alamo Bowie was by now sick and had taken to his bed, while Travis had taken complete command. Travis issued an appeal to the people of Texas, indicating that he was prepared to fight to the end. "I shall never surrender or retreat," Travis said, "I call on you in the name of Liberty, of patriotism and everything dear to the American Character . . . Victory or Death." It was an inspiring declaration and it must have lifted the morale of the men of the garrison, who were fighting for something more important than merely their own individual lives.

Meanwhile, Santa Anna's siege guns had begun a bombardment of the fort. Mexican sharpshooters picked off members of the garrison, but Crockett's Tennessee men were even more adept as snipers with their smooth-bore muskets, deadly at up to three hundred paces. As time passed it became obvious that no relief was coming to the Alamo and Santa Anna tightened his blockade. Texan morale was raised in some unusual ways, including Crockett's suggestion to use music. In fact, not only did the music raise morale among the defenders, but it also served to convince the enemy that their efforts were in vain. Scotsman John McGregor's bagpipes might have been enough to curdle the milk and Crockett's own fiddle might have set the cats fighting among themselves, but they spoke of the human spirit, unbreakable in spite of the parlous state the defenders were in. The sound of the music in the darkness was in itself the most powerful declaration of defiance.

On 1 March reinforcements at last arrived, limited as they were to the thirty-two men of Lieutenant George Kimball's "Gonzales Ranging Company of Mounted Volunteers." At least it showed Travis that he was not forgotten. But two days later the garrison received a setback. News was received that the expected help from Colonel Fannin at Goliad was not coming. It was a fatal blow and Travis realized that it was a death sentence for the Alamo and its garrison. With grim determination Travis set about the task of selling his life and those of his men so dearly that Santa Anna's victory "will be worse for him than a defeat."

At this moment legend has it that Travis drew a line in the dust and called on all those who were prepared to stay with him to defend the Alamo to the death to step across it. Melodramatically Bowie had his bed lifted across the line, and he was followed by most of the garrison. Did it ever happen? Does it matter? The fact is that Travis had set out his intention to die for his cause if necessary and the vast majority of his men chose to stay with him. It was courageous—foolhardy even—but it showed a fighting spirit that could not be quelled by the mere numbers of the enemy.

Santa Anna now had over two thousand men around the Alamo and he planned an offensive using the bulk of his men against the weaker north wall. At dawn on 6 March, 1836, the Mexican infantry advanced toward the Alamo. What happened inside the fort can only be gleaned from the Mexican sources, as none of the Texans survived. The courage of the defenders won the admiration even of their enemies. Travis, apparently, fought on the walls and, having just repulsed an assault by de Cos's column, he was shot through the forehead and died immediately. On the northern side, the Mexicans stirred the spirits of their men with a military band. The music, chosen for the assault by Santa Anna, was based on a trumpet call used in the fifteenth century by the Spaniards in their battles with the Moors. Its message was ominous: this was a fight to the death and no quarter would be given. It took three attacks to breach the northern wall, but then the Mexicans poured into the Alamo and a general massacre ensued. Crockett and his men, out of ammunition, fought with their backs to the old convent walls and used their muskets as clubs before they were all cut down and killed. In the old convent buildings, room-to-room fighting continued, during which Bowie was bayoneted, still in his bed but fighting with pistol and knife.

After a mere ninety minutes the battle was over. A few men had escaped the butchery by climbing the walls, but they were quickly cut down by the Mexican cavalry waiting outside the fort. Mexican losses had been severe: six hundred casualties in total of whom two hundred or more were killed. American sharpshooting had resulted in almost all the Mexicans being hit in the head or upper body. As one Mexican officer said, "The firing of the besieged was fearfully precise. When a Texas rifle was levelled on a Mexican he was considered as good as dead. All this indicates the dauntless bravery and the cool self-possession of the men who were engaged in a hopeless conflict." The Texans in the Alamo were not professional soldiers under orders to be there, or opportunists trying to gain something from the battle if they won. Instead, they fought with a freedom of spirit that was unusually American. They sacrificed their lives because they chose to do it, and fought for a cause that appealed to them. There was nobody behind them with a pistol in their backs; no sergeants carrying whips were needed to make them fight. They fought for freedom because they themselves were free. Some of them or their fathers had fought the British to win independence for their new nation and they were not prepared to submit to a paltry dictator like Santa Anna and his clockwork soldiers.

Revenge for the Texans came quickly. Inspired by the example of their martyred garrison of the Alamo, Houston's Texans defeated Santa Anna at the battle of San Jacinto just a few weeks later. If liberty and freedom were mighty motivators, so was revenge. Shouting "Remember the Alamo," the Texans tore into the Mexican army, killing over six hundred of them and capturing a further 730. Santa Anna was himself captured, but Houston resisted the urge to hang him and allowed him and the rest of his army to leave Texas. The new republic was recognized by the United States in the same year and ten years later became a member of the United States. The defenders of the Alamo had not only been avenged—they had been justified.

5

"The Thin Red Line" at the Battles of Balaclava and Inkermann, 1854–55

Writers who have dealt with the subject of fighting spirit have generally assumed that soldierly moral is—among other things—normally dependent on the way in which the soldier is treated. If he is neglected, badly clothed and fed, led into disaster by idiotic officers, and given the impression that he counts for rather less than the average saddle or bottle of fine wine, it is generally assumed that his combat performance will suffer. The British army in the Crimean War, however, provides an almost entirely contrary picture. Several times during the campaign in the Crimea, the British soldier revealed a spirit and a fortitude completely at odds with the abysmal treatment he was receiving from his own officers and administration. High morale was maintained in conditions that should have ensured discontent if not actual mutiny on the part of the common soldier. After all, nobody likes losing their luggage, but six weeks!

> The British soldier usually carries in his knapsack clothing sufficient to admit of at least one change, but on landing in the Crimea, so great was the anxiety to advance rapidly on Sebastapol, that authority was given to leave the knapsacks in the transports, and of this permission most of the corps appear to have availed themselves. Generally, therefore, each soldier had only a shirt, a pair of boots and a pair of socks, loosely rolled up in his blanket; and even of these a great part was lost at the battle of the Alma, or

during the march. This separation from the knapsacks could have produced but little inconvenience had it terminated with the march; but, unfortunately, they were left on board of the different vessels from which the men had disembarked, and these, being urgently required to bring reinforcements and supplies were sent to the opposite coast, carrying the knapsacks with them. It appears from the evidence that, on the average, more than six weeks elapsed before they were recovered, and then in many cases only after they had been plundered of a great part of their contents. The valises of the Cavalry, which were likewise left behind, shared the same fate, and the great majority of the troops were thus deprived, for a considerable time of a change of clothing.[32]

On campaign the welfare of British soldiers was the responsibility of the Commissariat, but so deficient was this organization in the field that complaints flooded back to London, with the result that in February 1855 Sir John McNeill and Colonel Alexander Tulloch were sent out to the Crimea to report. What they found was little less than a national scandal. As they wrote:

The deaths, including those at Scutari and elsewhere, appear to amount to about 35 per cent of the average strength of the army present in the Crimea from the 1 October, 1854, to the 30 April, 1855, and it seems to be clearly established that this excessive mortality is not to be attributed to anything peculiarly unfavourable in the climate, but to overwork, exposure to wet and cold, improper food, insufficient clothing during part of the winter, and insufficient shelter from inclement weather . . . much of the labour and exposure which the troops had to undergo was a consequence of the want of sufficient land transport, which it was the duty of the Commissariat to provide . . . In like manner, the injurious consequences of defective cooking were attributable to a deficiency of fuel, or of transport to convey

[32]G. Regan, *Someone Had Blundered,* p. 204.

it to Balaclava to the camp, both of which are matters affecting the Commissariat.[33]

The Commissioners were particularly concerned about the quality and quantity of the men's diet. It was obvious that the men were often reduced to half rations at short notice and sometimes received nothing. One example they cited concerned 25 December, 1854—Christmas Day, and even on campaign a traditional day of celebration in the British army—when Colonel Bell's command received no food at all. The colonel, quite reasonably, "kicked up a dust" with a commissary office and eventually some small portions of fresh meat were served out, but by that time it was dark and the men had no fires or means of cooking it. Salt meat and biscuit were generally available but these gave insufficient nutritional value to men working such long hours and in such terrible conditions. A lot of the soldiers could not eat the salt beef even when full rations were issued because "the prevailing diseases were . . . affections of the bowels, in most cases connected with scorbutic tendency in the system; and not only the men, but the medical officers also, believed that those diseases were aggravated . . . by the continued use of salt meat." As a result, much salt meat was simply thrown away, sometimes hundreds of pounds a day by some regiments. Why more fresh meat was not made available is difficult to understand, especially when there was no shortage of cattle available in neighboring states; in December 1854, when only salt meat was issued, the Commissary General admitted to having eight thousand head of cattle secured.

McNeill and Tulloch were particularly critical of the way that salt meat damaged the men's health. It was well-known at the time that where large armies were encamped there was usually a large number of men suffering from diseases of the bowels, and it was necessary that every precaution should be taken to prevent this. Experience had proved that the use of salt meat in the diet increased both the

[33]G. Regan, *Someone Had Blundered,* p. 196.

number of those attacked by scurvy and the number of fatal cases. If one added to this the unusually difficult circumstances of the troops, the heavy work, the long hours, the winter season and the inadequate clothing and shelter, it was absolutely vital that "no practicable means ought to have been left untried to protect the troops from the injurious effects of diets—one of the few conditions of the soldier's existence which were absolutely within control."

Clearly lives were being thrown away through sheer bureaucratic incompetence. As I have written elsewhere, "It would have been more trouble for the Commissariat to issue fresh meat than salt meat, to bake fresh bread rather than distribute bags of biscuits, but it would have saved the lives of many and relieved the sufferings of many more. Perhaps it was impossible for the civil service mentality to come to terms with the horrors that surrounded it, and it accordingly took refuge in its own ordered and regulated world, where casualties were just numbers on a sheet of paper." Research has shown that the Crimean soldier was receiving food of less nutritional value than inmates of British prisons! Nor, although lime juice was available in great quantities aboard one of the ships in Balaclava harbor, was any of it circulated to the troops, even though many of them were succumbing to scurvy. Coffee beans were distributed to the men, but only in their raw, uncooked form. As one officer remarked, "A ration of green raw coffee berry was served out; a mockery in the midst of all this misery. Nothing to roast coffee, nothing to grind it, no fire, no sugar; and unless it was meant that we eat it as horses do barley, I don't see what use the men could make of it, except what they have just done, pitched it into the mud." In fact, the British soldiers were more resourceful than that. Some of them ground the coffee with cannon balls and cut their dried meat into strips to make fuel to roast the coffee beans.

As well as poor food, the British soldiers had been equipped with unsuitable clothing. The main problem was their boots, which were inadequate for the harsh conditions of a Crimean winter. In the snow men often wore two pairs of socks, while others suffered from swollen feet from long periods of duty in the soggy trenches. As a result the boots ordered from England by the Commissariat to replace those originally issued, which had worn out, were far too small for the men. Many had to wear boots so small "that women could hardly have got them on." Consequently, the British soldiers took to stripping the excellent Russian boots from corpses on the battlefield. The British boots were so badly made that the following curious incident seemed to symbolize the entire Commissariat effort in the Crimea. On 1 February, 1855, the 55th Regiment paraded in "a vast black dreary wilderness of mud." The men sank into the slime, and, as they tried to march away, all the boot soles were sucked off. The men simply threw away the boots and marched off to the front in their socks! That was fighting spirit defined by a gesture.

It is against this background that one needs to set the combat performance of the British army during the Crimean War. What is ironic is that the confidence of the soldier in his own ability, and that of his regiment, was higher perhaps at this time than at any other in British history. It seemed that the worse his treatment, the better the British soldier fought. Certainly both the battles of Balaclava in 1854 and Inkermann in 1855 contain epic scenes of courage and fortitude that have inspired later generations in their own wars.

Between Waterloo in 1815 and the outbreak of the Crimean War in 1854, the British army had enjoyed a period of genteel atrophy. The army mirrored the aging of the duke of Wellington himself. Unchallenged by any European foe, it had confined itself to colonial conflicts in which its superior firepower, even though allied to outdated tactics, had usually sufficed to provide victory. Such unchallenged mastery had fostered an overweening feeling of superiority that was at odds with reality. The British soldier—and even more his commander—felt that it was enough to turn up for the enemy to quit the field in rout; the shades of the "Iron Duke" would see to the rest. And it was with this attitude that the British entered the war against Russia in the Crimea with the French as their allies. The more progressive French regarded the British as something from another—simpler—era, where courage was enough and the conduct of war could comfortably be left in the hands of gentlemen. Even the Russians found the British tactics unconventional to say the least and positively idiotic on occasions. Whatever foreign opinion was on the subject, the British were disinclined to change their tactics. After all, they had been good enough for the duke of Wellington in Spain and—and this is the main point—they still produced victories.

As if to prove the point, in the first battle of the war the British clambered up the slopes to the Alma Heights in line formations nearly a mile wide and devastated the Russian defenders with their massed firepower. One of the Russian officers was astounded enough to comment that he could hardly believe it possible "for men to be found with sufficient firmness of morale to attack, in this apparently weak formation, our massive columns." Another Russian was surprised at "the leisureliness of their gait." The confidence of the British soldier and his fighting spirit was far greater than that of the Russian conscripts he faced. Although heavily outnumbered, the British soldier possessed sublime confidence—often based on a total ignorance of what he was facing—in the superiority of his regiment, his comrades, and himself over any foreigners. This insularity, indeed chauvinism, was to sustain the redcoat on battlefields throughout the nineteenth century until the truth confronted him at Mons in 1914.

In spite of the solidarity of the British regiments throughout the Crimean War, the soldiers were individuals and felt individually. It was all very well for officers to generalize about the "Bull-dog courage" of the troops, but men like Private Hyde got closer to the truth when he wrote of his own experiences: "Whether it was fear or excitement I don't know, but I seemed dazed, and went wherever the others went, and did what they did; there was nothing to be gained from hanging back." The slaughter on the Alma had been terrible. Even though the Russians had been put to flight the battle had cost the British 362 dead and 1621 wounded.

One feels almost sorry for the French and Russians during the Crimean War, trying to fight a modern war, but constantly upstaged by the British, who assumed the starring role in tableau after tableau. At Balaclava, where the Russian cavalry had taken the sleepy British commanders completely by surprise and were about to overrun the British guns, Sir Colin Campbell entered on cue with the 93rd Highlanders, to make his contribution to British military history: "the thin red line." The 93rd—just 550 Highland soldiers—covered a front of about 150 yards. Under constant fire from Russian artillery they faced a mass of Russian hussars bearing down on them. Alongside them Turkish troops had panicked and fled into the British camp in disorder, attacked as they passed by one of the Scotsmen's brawny wives. As they bore down on the line of Highlanders the Russians were astonished to find that the 93rd did not form up the "square," the traditional defense of infantry against cavalry. Instead they stood their ground in just two lines, bringing all their muskets into play simultaneously. As they waited, Campbell's voice could be heard: "There's no retreat from here, men! You must die where you stand." Again group solidarity was the answer. The men of the 93rd would live and fight and die together. Private John Scott replied to Campbell, "Aye, aye, Sir Colin, and needs be we'll do that." There is no suggestion that the nineteenth century soldier valued his life any less than we do today. He merely seemed to be able to suppress his own needs—at least temporarily—in terms of a greater or higher duty. "King and Country" may have a faintly ridiculous air to it today, but for men like John Scott at Balaclava and thousands like him it had real meaning, giving purpose to a life that might otherwise seem cheap and mean. "Fight or flight" was clearly operating at this moment of tension. Standing still and merely holding the line in the face of cavalry attack and artillery bombardment was almost more than the men could bear. As their Turkish allies fled some of the Scotsmen began to grow restive, as if they wanted to charge the Russians rather than wait to receive them. But Campbell was up to his task—one of the few British officers of his day who really was—and snapped at them: "93rd! 93rd! Damn all that eagerness." As the Russian hus sars broke into a canter and then a charge, the 93rd fired a first volley and then calmly reloaded. A second calm volley at 250 yards brought many riders tumbling down. And now the hussars began to sheer away from their indomitable opponents. But the 93rd was not finished with them. As they rode towards the right flank of the line the Grenadier Company of the 93rd hit them with a third shattering volley that sent them racing away in disorder. The British camp was saved as the Highlanders tossed their feathered hats into the air. Regimental spirit had triumphed once again.

The strength of the British soldier has long been associated with the feeling of *esprit de corps* which he gained from membership of the regimental system. Certainly this carried him through the initial battles of the Crimean War, at Alma and Balaclava. But the fighting at Inkermann seemed to contradict this. For a start it was one of the most confused battles of the nineteenth century, a true "soldier's battle" in which the role of the commander devolved upon subalterns as a result of the close fighting and the prevailing fog. The larger formations seen elsewhere were reduced by circumstances down to platoon and company level. The enemy were everywhere, it seemed, in front, behind and frequently all around the heavily outnumbered British soldiers. Even the superior fire discipline of the British and the greater hitting power of their Minié rifles were rendered nugatory by the close-range fighting. Much of the fighting, indeed, resembled more a huge melee in which muskets were used as clubs and men wrestled each other to the ground.

In such a melee self-preservation asserts itself against feelings of pride of regiment or race. Frequently outnumbered ten to one the redcoats fought with unimaginable savagery. As Private Hyde of the 48th later said,

> They came on like ants; no sooner was one knocked backwards than another clambered over the dead bodies to take his place, all of them yelling and shouting. We were not quiet and what with cheering and shouting, the thud of blows, the clash of bayonets and swords, the ping of bullets, the whistling of shells, the foggy atmosphere, and the smell

> of powder and blood, the scene where we were was beyond the power of man to imagine. We had to fight to save our lives.[34]

Individual courage in such a battle was casual, yet often decisive. Conducted in a fog many gallant deeds passed unnoticed. Yet the fog was vital to the British, by concealing from the Russian masses how few were their opponents. Given one glimpse of the truth the Russians would have taken heart and swept away the few redcoats who stood in their path. Small groups of soldiers wandered about lost in the swirling fog. One group of Coldstream Guards spotted Sir Charles Russell and called out to him, "If any officer will lead us we will charge." No blue-blooded British aristocrat could resist such an invitation. Russell went over to them and said, "Come on, my lads, who will follow me?" With a cheer the guardsmen followed Russell down into the Russian masses. In the confused fighting that followed Russell suddenly felt a gentle tap on his shoulder. He turned to find Private Anthony Palmer looking at him. "You was nearly done for," said Palmer. Apparently a Russian had been about to bayonet Russell in the back when Palmer saw the danger and killed the Russian with his rifle butt. Another of Russell's followers, Private Bancroft, was in the heart of the struggle:

> I bayonetted the first Russian in the chest; he fell dead. I was then stabbed in the mouth with great force, which caused me to stagger back, where I shot this second Russian and shot a third through. A fourth and fifth came at me and ran me through the right side. I fell but managed to run one through and brought him down. I stunned him by kicking him, whilst I was engaging my bayonet with another. Sergeant-Major Alger called out to me not to kick the man that was down, but not being dead he was very troublesome to my legs; I was fighting the other over his body. I returned to the Battery and spat out my teeth; I found two only.[35]

[34]M. Barthorp, *Heroes of the Crimea,* p. 95.
[35]Ibid., p. 103.

British phlegm had triumphed over Russian stoicism. Yet elsewhere, a few hundred British troops—mainly guardsmen—led by Sir George Cathcart and the duke of Cambridge at Sandbag Battery were virtually surrounded by three Russian battalions. In the immortal words of Sir George, just before he was killed, "We are in a scrape." Their position seemed hopeless and men began to waver. One sergeant called out that they needed "the greatest miracle in the world." Meanwhile, in a lull, one of the young drummer boys, just ten years old, had brewed up some tea. It had its effect. Assistant Surgeon Wolseley, returning from helping some wounded, found himself blocked by hundreds of Russian soldiers. Undeterred, he rounded up some guardsmen from near at hand. As he went on, "I was the only officer in sight and gave the order, 'Fix bayonets, charge, and keep up the hill.' We charged through losing, I should think, about half our number." This charge completely unsettled the Russians, who in the fog could not tell if it presaged an even larger assault. They fell back, allowing Cambridge and his guardsmen to wriggle out of their trap. Captain Higginson remembered the desperate moments when it seemed that the Grenadier Guards would lose their colors:

> Clustered round the Colours, the men pressed slowly rearwards, keeping their front full to the enemy, their bayonets ready at the "charge." As a comrade fell, his fellow took his place and maintained the compactness of the diminishing group that held on in unflinching stubbornness. More than once from the lips of this devoted band came the shout, "Hold up the Colours!," fearing no doubt they might lose sight of those honoured emblems. The two young officers, Verschoyle and Turner, raised them well above their heads and in this order we slowly moved back, exposed to a fire, fortunately ill-aimed, from front, flank and rear.[36]

[36]M. Barthorp, *Heroes of the Crimea,* p. 103.

Elsewhere, Captain Burnaby of the Grenadier Guards charged nearly seven hundred Russians with just twenty of his men and put them to flight.

The Russians could not have been aware that they were involved in a battle that they could not win. They were fighting more than British soldiers; they were fighting British history and regimental tradition. In the terrible bloodsoaked encounters in the surreal atmosphere of that foggy day, they were up against regiments who counted Minden and Albuera on their list of battle-honors. And the men remembered the spirit with which those other men—those regimental heroes of yesteryear—had fought in similar impossible actions and were inspired by the knowledge that they had triumphed over the odds. Were the men of their day less worthy? From the 20th Regiment welled up the "unearthly Minden yell" as they drove the Russian Iakoutsk Regiment at bayonet point towards a ravine. From another part of the field the officers of the 57th—the "Diehards"—enjoined their men to "Remember Albuera!"

The chaos was tremendous. Men fought singly, or in small groups, though rarely with the comfort of knowing that their usual comrades were at their side. Regimental pride gave way to personal pride. Men were fighting, unseen by all those who mattered to them. No fear of personal shame or failure kept them in the fight. Many could have run away with perfect security, yet few did. The bloodletting had taken on the form of an enormous struggle in which everyone must take part. It was not sport—it was too bloody for that—yet it was as thrilling as any sport could be. Men found themselves confronted with situations they would never again face. No drillmaster could have contrived such challenges. No commander would have led them into places where the enemy outnumbered them ten to one. No officer would have led the charges that they led that day without a second thought, had the excitement of the struggle not been in his blood. Later men wrote rationally about these events when their minds had cleared and their bloodlusts had drained away into some more appropriate form. But for a few hours at Inkermann, the British soldier enjoyed the forbidden pleasures of slaughter and mayhem, without responsibility.

Historians have found patterns in the action at Inkermann, yet there were few that made any sense to the common soldiers. They fought for love: love of life, love of their friends, their regiment, and their Queen and Country. And those who survived the day—like those who lived through the fighting at Waterloo—felt that life had little more to offer. An officer who missed the battle spoke—nose pressed to the window of these men's experience—of what he thought had happened at Inkermann:

> We owe our existence to the pluck of the private soldiers. Surprised by immensely superior forces, without orders, without reserves, sometimes without ammunition when, after using that in the pouches of the dead and dying, they pelted the enemy with stones. Mixed up in strange groups away from their comrades they knew and the officers they trusted, the men held the ground they stood on as long as life was in them. Two mounds mark where hundreds of our dead have been packed away together—"worthless refuse, what parents have cherished, friends esteemed and women loved."[37]

But he, of course, had missed the battle. The "refuse" would not have agreed. For them was the true meaning of Horace's phrase: *Dulce et decorum est pro patria mori.* The reporter from the *Times,* W.H. Russell, summed up the historical significance of the battle and of the fighting spirit of these incredible men.

> It is considered that the soldiers who met these furious columns of the Czar were the remnants of three British divisions, which scarcely numbered 8,500; that they were hungry and wet, and half-famished; that they were belonging to a force which was generally "out of bed" four nights out of seven; which had been enfeebled by sickness, by severe toil, sometimes for twenty four hours at a time with-

[37]M. Barthorp, *Heroes of the Crimea,* p. 143.

> out relief of any kind; but among them were men who had within a short time previously lain out for forty eight hours in the trenches at a stretch—it will be readily admitted that never was a more extraordinary contest maintained by our army since it acquired a reputation in the world's history.[38]

Prosperity was the enemy of the Crimean veteran's patriotism. As late-Victorian Britain purged herself of her industrial sins, like poverty, squalor and disease, such men no longer needed to enlist in the army. They could find "better work" in factories and mines and, as they grew prematurely aged, they could dream of forbidden pleasures.

[38]M. Barthorp, *Heroes of the Crimea,* p. 143.

6

The 20th Maine at Little Round Top— The Battle of Gettysburg, 1863

The battle of Gettysburg has a rare quality that belongs to few other battles—perhaps Waterloo and Balaclava only—in that the actions of the participants are lifted out of the merely ordinary and achieve an epic, almost-legendary nature. While at Waterloo, neither Napoleon nor Wellington had their best troops available, yet in a hundred parts of the field men exceeded themselves so completely that the vile process of killing and being killed achieved a beauty that was born of human endeavor as much as human suffering. So at Gettysburg, the combatants on both sides seemed to possess a deep fund of fighting spirit into which they were able to dig deeper than ever before. Perhaps "Johnny Reb" never fought so well again except in desperation. Much of the confidence that had previously driven him forward against the whole world if necessary perished on Cemetery Hill, while the superiority that he had always assumed was his over "Billy Yank" received a stinging rebuttal on Little Round Top. Not only had General Robert E. Lee shown himself vulnerable at Gettysburg, but the moral advantage that the Rebs held up to that point began its inexorable slide. The combat performance of the 20th Maine at Little Round Top on 2 July, 1863, was one of those examples of men mastering their fears and achieving a unity of action that one might have called sublime.

The Union army commander, General George Meade, had not prepared the Gettysburg position for a defensive battle in advance, as Wellington had at Waterloo in 1815. When fighting broke out there on 1 July, it was more of an encounter battle than the product of careful planning. As a result, it is possible to forgive Meade for not immediately seeing the strategic importance of the two "Round Tops" to the left of Cemetery Ridge. However, it is surprising that, by the second day of the fighting, they were still bare of Union troops, except for some signalers. It is to the eternal credit of General Gouverneur K. Warren, an engineer by profession and a man with a keen eye, that he followed his own curiosity and investigated what was going on beyond the two hills for himself. What he found there was of deep concern. To his great surprise Warren found them undefended. What was worse, the woods facing them could offer cover to any advancing Confederate troops. He made the sudden decision to test out one of his fears. He ordered a gun below him on a smaller hill called Devil's Den to lob a shot into the woods just in case . . . As Warren watched the woods, the unmistakable glint of sunshine on bayonets was revealed, as the Confederates watched the artillery shell loop into the sky. His worst fears were confirmed. The Rebels were about to outflank the entire Union position on Cemetery Hill. They would take Little Round Top and from there command the entire battlefield, rolling up Meade's army in short order. But why were they waiting with such an opportunity in view? Warren snapped into action. The only answer was to get some troops onto Little Round Top before Longstreet—for it was he commanding the Confederate right—swept into action. Warren sent a rider to request a division from Meade; he also sent to General Sickles, commanding III Corps, asking for a brigade. But Sickles was already in enough trouble against Longstreet's men and could spare nobody. Fortunately for Warren—and the Union cause—Meade had ordered General Sykes's V Corps to advance to the support of his left wing. The leading brigade of Sykes's corps was led by Colonel Vincent, and marching with it was the 20th Regiment of infantry, the Maine Volunteers, for their date with destiny.

As Vincent's brigade arrived at Little Round Top they

were brought under fire from the Confederate artillery. With shells stripping the trees of branches and fragments of rocks flying everywhere, Vincent coolly brought his men into formation, with Colonel Chamberlain's 20th Maine regiment holding the far left of the entire Union line, bent round Little Round Top but just below the crest. With no support whatsoever on his left, Chamberlain sent Captain Morrill at the head of Company B to cover his exposed flank and to act as a reserve in case of emergencies.

On the Confederate side, it had taken some time for the generals to realize that the two hills—the Round Tops—on the left of the Union position had not been fortified. Surely that was too good to be true. Was tortoise George Meade trying to trap the Rebel hares? It was so obvious that the hills had to be fortified that it was just possible that nobody had actually got round to doing it. In any case, those hills were the key to the whole battlefield and the Confederates decided that they must take them. But the movement of military units on maps was far too easy for generals. For the men on the ground it always meant hard marching, and although the Confederate army was one of the greatest marching armies in history, the twenty-eight miles that General Law's brigade of Alabamians had covered since 3:00 A.M. had sapped their energies. Moreover, before the men were able to get a drink and refill their canteens, they were marching into action. Even for men with such fighting spirit as the Rebs it was asking a lot. Inevitably, combat efficiency declines as the mind becomes increasingly concerned with the body's physical condition. Nevertheless, the military situation could take no account of the condition of the men. It was a race now for a position that could decide the battle and, in the minds of all Southerners, possibly decide the outcome of the war. Little Round Top was a target worth thirsting for. The Alabamians, already footsore, now found themselves under a galling fire from Union sharpshooters on Devil's Den and were beginning to have to stumble and climb over the rock-strewn sides of Big Round Top. Men were fainting in the heat and others were falling out of line with exhaustion. Colonel Oates, commanding the 47th Alabama, occupied the crest of Big Round Top and ordered a halt for the water-carriers to catch up. But the men sent to refill the canteens had walked into an ambush and no water was to reach the fighting Alabamians. Nevertheless, on Big Round Top Oates began to see how he could bring up artillery and completely undermine the Union position. Oates wanted to fortify the position, but he was ordered to come down and turn the flank of the Union army on Little Round Top. This was a decisive opportunity that the Confederates would miss like so many in this battle, but in the heat of the action it is often difficult to establish priorities.

Meanwhile, back on Little Round Top, the 20th Maine was waiting for what they knew would be a tremendous struggle. Fully aware of the strategic significance of their posting, every man must have been assailed by the twin feelings of fear and pride. To fight and perhaps to die in battle was one thought, but to have the experience given an added significance was enough to boost morale. The importance of the task allocated to them must have lifted every man's fighting spirit. In the words of John J. Pullen, historian of the regiment:

> These minutes of inactivity would be almost intolerable, but blind instinct would be getting their bodies ready—blood beating harder and faster through the arteries; lungs starting to dilate deep down, reaching for more oxygen; stomach and intestines shrinking and stopping all movement; and tension rising to the point where it could shake a man like the passage of a powerful electric current. When it came, any kind of action would be a relief—and the reaction would be explosive.[39]

Standing on the extreme left of their brigade, the 20th Maine could hear the battle around the hill developing as each regiment on their right came into action one after another against the Confederates as they marched around the Union flank. Soon bullets were hitting their own men. But even now they could not fire. The enemy was not yet

[39] J. J. Pullen, *The Twentieth Maine,* p. 123.

in sight. Tension reached a critical point. All the sounds of warfare—the sounds of muskets being loaded, and fired and men crying out, and shouting and screaming—were present but as yet none of the sights. Then came the Rebel yell and—at last—the foremost troops of Law's brigade came into sight. They were the 4th Alabama, and as they charged up the slopes of the hill, dodging rocks and boulders and vaulting over tree stumps and undergrowth, the 20th Maine opened fire. No sooner had the first volley been fired than a crisis was at hand and it fell to Colonel Chamberlain to solve it. An officer brought up the unwelcome news that a Rebel column of men was already *behind* them. Screened by the smoke of gunfire Colonel Oates and his Alabamians—who had come down from Big Round Top—had completely outflanked Vincent's brigade. Moreover, Oates had two regiments with him—15th and 47th Alabama—and outnumbered Chamberlain by two to one. It was a crisis that needed the attention of the calmest of men. Fortunately, it received it. Colonel Chamberlain—a professor in civilian life—weighed up the problem in a moment and ordered his regiment to reform. What occurred is complicated to explain, let alone enact in the heat of battle and under heavy enemy fire. The morale of the 20th Maine and their confidence in their officers must have been of a supreme nature to enable them to respond as they did. The shades of Prussian drillmasters of the eighteenth century would have nodded their approval. In the words of John Pullen:

> The plan was executed in a way that never thereafter ceased to be a source of wonder to the officers of the 20th Maine. With bullets smashing into it, and the roar of gunfire making commands inaudible, the regiment writhed and twisted into the new formation like a single, living organism responding to a sense of imminent danger. Or—it was almost as though every man had been party to a quiet conference, where everything had been diagrammed and perfectly understood. On the right wing, men were firing, shouting dodging from rock to rock and tree to tree, and gradually forming a single rank that covered the entire original front

of the regiment. Chamberlain remembered that while this was going on there seemed to be no slackening of fire on the front.

Meanwhile, the men of the bent-back wing were forming a solid line facing to the left, taking what concealment they could find behind rocks and undergrowth. Their presence came as a grievous surprise to the 15th Alabama.[40]

On the slopes of Little Round Top, Oates's Alabamians could sense victory. Above them was the unguarded flank of the entire Union army. Just as the 20th Maine had felt the honor of guarding the post of greatest danger, so the Alabamians felt the elation that comes with the knowledge that while many thousand of your colleagues are toiling behind you and exchanging volleys with the enemy, to you has fallen the prize—the point of penetration, the stroke that will win the battle and, perhaps, earn undying glory by winning the war itself. If the fighting spirit of the 20th Maine was at its height, that of the 15th Alabama can hardly have been much lower. For a moment, the tiredness of hours of long marching and the thirst they all felt would have been put aside as their bodies primed themselves for action. At last, giving the eerie Rebel yell, the Alabamians stormed up the slope of the hill, heading for glory. Rarely can hopes have been so rudely dashed. In mid-flight the Rebels were hit by a volley from men who rose as if from the ground to deliver their fire. It was delivered at short range and it was devastating. The shock and surprise of finding what had been assumed to be an unguarded crest defended must have damaged much of the confidence the Alabamians had felt. Their charge stopped as if they had met a brick wall. For a moment they wavered. But then their spirit revived and they remembered how close they were to victory. They went on and drove into the Maine volunteers at bayonet point. The crisis was at hand. In the chaos nobody had given the order for the Maine men to fix bayonets. In the hand-to-hand fighting the advantage had to be with the Rebels. But Chamberlain's men were

[40] J. J. Pullen, *The Twentieth Maine,* p. 118.

inspired and fought with unprecedented ferocity. It was almost too closely packed for bayonets and men used their muskets like clubs or even wrestled or fought with their fists. Men felt a bloodlust which had nothing to do with their training. It was primitive and instinctual. It had been present in every battle that mankind had ever fought. It was not just the other side of self-preservation. It was a kind of impulse to kill which overthrew all cultural imperatives, all Christian conditioning. It was a frenzy in which men killed or were killed. There was no time for thought and so taking prisoners was almost impossible. Mercy required time to think and in the chaotic melee on Little Round Top few men had time to think. But one thing was becoming evident to Colonel Chamberlain: his men were holding their own against vastly superior numbers.

Individual moments of drama passed before Chamberlain's eyes. One boy, hideously cut across the head and sent to the rear, presumably to die, was next seen fighting furiously in the front line with a rough bandage round his forehead. Two men, put under close arrest for mutiny, chose this moment to break away from their confinement and join the battle, earning an instant pardon. Private Theodore Gerrish, who later wrote an account of the fight, can be excused some license in his description, "Not only on the crest of the hill, among the blue coats, was blood running in rivulets and forming crimson pools, but in the gray ranks of the assailants there had also been a fearful destruction."

The men of the 20th Maine had started the battle with sixty rounds each, but within an hour they were running short of ammunition. John Pullen estimates that they must have fired twenty thousand bullets in that time and, assuming that the Alabamians returned almost as many, it is easy to understand the astounding damage to trees on the hill. Every tree had been ripped and gashed by gunfire to a height of six feet, and several had been felled by bullet fire alone. Up and down the slope the battle swayed. It was one of the fiercest small actions of the entire war. Colonel Oates later recalled that his men overran the Union position five times but were thrown back each time by counterattacks.

Oates eventually drew back his regiments to re-form and this gave the men from Maine a chance to breathe and assess their losses, which amounted to at least a third of their number. But the ammunition situation was now desperate and the Alabamians below still held a numerical advantage. Eventually, Chamberlain reasoned, they would overrun his men by sheer weight of numbers. He therefore concluded that his men could get no help from his neighboring regiments, who were themselves deeply involved with the enemy, nor any more ammunition, nor indeed could they fall back to another position. If he could not go back and he could not stand still, that left just one course of action. Chamberlain ordered his men to fix bayonets and prepare to charge Colonel Oates's men. Both sides had displayed courage enough for ten battles. What was needed now was self-belief and willpower. Chamberlain decided to impose himself on the enemy. Charging down gave his men momentum, as well as the moral advantage of action rather than passivity. Charging spoke of belief and confidence and, if nothing else, a frightening desperation. Men who would come down off the crest of the hill had come to fight and to overcome. All these thoughts must have sapped the spirit of the Alabamians, rendering them susceptible to the thirst they must have been feeling by now and the overwhelming exhaustion that would assert itself as the initial excitement of the struggle began to abate. Private Gerrish described what happened next.

> A critical moment had arrived, and we can remain as we are no longer, we must advance or retreat. It must not be the latter, but how can it be the former? Colonel Chamberlain understands how it can be done. The order is given "Fix bayonets!" and the steel shanks of the bayonets rattle upon the rifle barrels. "Charge bayonets, charge!" Every man understood in a moment that the movement was our only salvation, but there is a limit to human endurance, and I do not dishonor those brave men when I write that for a brief moment the order was not obeyed, and the little line seemed to quail under the fearful fire that was being poured upon it. O for some man reckless of life, and all else save his country's honor and safety, who would rush

> far out to the front, lead the way, and inspire the hearts of his exhausted comrades!
>
> In that moment of supreme need the want was supplied. Lieut. H.S. Melcher, an officer who had worked his way up from the ranks, and was then in command of Co. F, at that time the color company, saw the situation, and did not hesitate, and for his gallant act deserves as much as any other man the honor of the victory on Round Top. With a cheer, and a flash of his sword, that sent an inspiration along the line, full ten paces to the front he sprang—ten paces—more than half the distance between the hostile lines. "Come on! Come on! Come on, boys!" he shouts. The color sergeant and the brave color guard follow, and with one wild yell of anguish wrung from its tortured heart, the regiment charged.[41]

So sudden was Chamberlain's charge—and, perhaps, so unexpected, in view of his heavy losses—that the Alabama men had no chance to fire a volley. They were simply rolled over by the momentum of the 20th Maine. For a moment they wavered, as if to resist, and then the bubble was burst—their confidence simply fell away. One of the Confederates fired at Colonel Chamberlain from close range but missed. As if symbolizing the entire struggle, this miss was virtually the last resistance the Alabamians showed on Little Round Top. The officer in question promptly surrendered to Chamberlain, while the rest of his men simply turned and ran. The collapse was total and irreversible. Their courage, always more fragile than the 20th Maine's, through thirst and exhaustion, simply ran out.

At the bottom of the hill, Colonel Oates received reinforcements from the 4th and 5th Texas regiments and even at this last moment it was possible that the outcome of the battle could be affected. But Captain Morrill and his Company B, sent out at the start of the battle to guard the regiment's flank, chose this moment to intervene. Morrill's men had been hidden behind a stone wall and they suddenly rose up and fired a volley into the backs of Colonel

[41] H. S. Commager, *The Blue and the Gray, Volume 2,* pp. 620–21.

Oates's men. It was the final straw: they were now being attacked from front and rear. Men could not be expected to stand any longer when they feared that they were surrounded. Panic was already beginning to spread in the ranks of the Confederates. It has to be admitted that Oates's leadership at this moment was panicky. He issued the order for his regiment to try to cut its way out of what he believed to be a trap. As he said:

> I had the officers and men advised the best I could that when the signal was given that we would not try to retreat in order, but everyone should run in the direction from whence we came . . . When the signal was given we ran like a herd of wild cattle, right through the line of dismounted cavalrymen . . . as we ran, a man named Keils, of Company H, from Henry County, who was to my right and rear, had his throat cut by a bullet, and he ran past me breathing at his throat and the blood spattering.[42]

The last observation by Oates, a minor point in so bloody an encounter, seems strangely to have stuck in his mind. It probably indicates that he was a prey to his own fears at this moment and was thinking of himself and of other individuals rather than of his command. In a sense, he had cracked and was incapable of further leadership. Aware of this, his men ran from their own individual fears, the strength of their group morale shattered.

With the Confederates in flight—400 prisoners were eventually taken—the 20th Maine returned to the crest of Little Round Top to assess the situation. The sight that met their eyes was a bloody one. They had begun the fight with 386 men, including, according to Colonel Chamberlain, "every pioneer and musician who could carry a musket." In total they had suffered 40 men killed and a further 130 seriously wounded. Amidst the carnage on the hilltop there were more than 150 dead and wounded Confederates. Although they could not have known it, the regiment had

[42] J. J. Pullen, *The Twentieth Maine,* p. 126.

fought one of the most important actions in American history. As Corporal William T. Livermore later wrote:

> The Regiment we fought and captured was the 15th Alabama. They fought like demons and said they were never whipped before and never wanted to meet the 20th Maine again . . . Ours was an important position, and had we been driven from it, the tide of battle would have been turned against us and what the result would have been we cannot tell.[43]

[43]J. J. Pullen, *The Twentieth Maine*, p. 127.

7

"The Legend of Johnny Reb"— The Battle of Missionary Ridge, 1863

"Johnny Reb" stands high in the pantheon of history's toughest fighters, yet for all his attacking qualities—which some writers have referred to as his "Celtic fury"—he was sometimes fragile in adversity and lacked the defensive qualities of his adversary "Billy Yank." As many Confederate officers complained throughout the Civil War, the Confederate soldier was impulsive and inclined to panic when matters took an unexpected turn. The result was that for all his marching skills—and they were truly formidable—"Johnny Reb" was sometimes confused over direction when it came to running. Nor was this merely a reaction to war weariness and defeat as the war drew to a close. From the very start there was a substantial proportion of cowards, fainthearts, and deserters in the Confederate ranks. This demonstrates the dangers of generalizing about the fighting qualities of group stereotypes.

Johnny Reb knew what he was fighting for. His officers told him often enough. Before the battle of Shiloh in 1862, General Albert Sidney Johnston appealed to his men:

> With the resolution and disciplined valor becoming men fighting, as you are, for all worth living or dying for, you can but march to a decisive victory over the agrarian mercenaries sent to subjugate and despoil you of your liberties, property, and honor. Remember the precious stake involved; remember the dependence of your mothers, your wives, your sisters, and your children, on the result; remem-

ber the fair, broad, abounding land, the happy homes and the ties that would be desolated by your defeat.[44]

But this sort of inspirational propaganda was lost on some men when the bayonets flashed and the bullets flew. Every army of any size has its fair share of cowards, men who will succumb to their private terrors and seek every opportunity to escape personal danger or responsibility. And the Army of the Confederacy was no exception. Southern generals wrote of this problem in astonishment as if they alone were subject to desertions and cowardice. This was far from the truth. Blue and gray shared the dangers of conflict fairly equally. There were heroes on both sides just as there were cowards. The difference, perhaps, is the legendary "courage" or fighting spirit that is associated with the soldiers of the South. At Gettysburg, it is Pickett's Charge that is most often remembered, not Chamberlain's Charge at Little Round Top. Pickett's men marched at the order—the flawed order—of their commander; Chamberlain's men charged because their blood was up and they were following inspirational officers. Pickett's Charge lost the battle for Lee; Chamberlain's Charge won the battle for Meade.

In his pioneering study of the Confederate soldier, *Johnny Reb,* Bell Irvin Wiley draws attention to the deliberate concealment of cowardice and desertion by regimental officers. He cites an example of the 31st North Carolina Regiment, in an action at Morris Island. The regimental commander had described their action as "cool and brave." However, the divisional commander saw things rather differently, commenting that, "The Thirty-first North Carolina could not be induced to occupy their position, and ingloriously deserted their ramparts . . . I feel it is my duty to mention their disgraceful conduct." Clearly the military historian needs to beware in reading the regimental reports where the reputations of officers as well as their men are involved.

[44]B. I. Wiley, *Johnny Reb,* p. 68.

In 1862 General D. H. Hill wrote, "Several thousand soldiers . . . have fled to Richmond under pretext of sickness. They have even thrown away their arms that their flight might not be impeded." At the battle of Seven Pines, Hill says "a few regiments disgracefully left the battlefield with their colors." General Whiting records that at Gaines's Mill, "Men were leaving the field in every direction and in great disorder . . . men were skulking from the front in a shameful manner; the woods on our left and rear were full of troops in safe cover from which they never stirred." One might inquire at this point what their officers were doing to force the men back to their positions? Clearly, even this early in the war, the fighting spirit of the Southern soldier varied considerably from regiment to regiment. At the battle of Malvern Hill, Confederate general Jubal Early noted, "a large number of men retreating from the battlefield." He personally saw, "a very deep ditch filled with skulkers," and found "a wood filled with a large number of men retreating in confusion."

In spite of Albert Johnston's attempts to lift their morale, many Rebels skulked or fled from the battlefield at Shiloh. One entire Tennessee regiment broke and ran back through its supports, scattering them to the winds and spreading panic in the rear with their cries of "Retreat! Retreat!" At least on this occasion their officers managed to rally them, but not for long. No sooner had they resumed their march toward the Union lines than their terror overtook them and they broke for a second time, trampling the color-bearer of the regiment behind them. The same thing happened to a Texas regiment at Shiloh which, having fired a single volley, broke and ran from the field. When one of their officers tried to rally them and called them "a pack of cowards," they replied that they "did not care a damn what they were called but they would not follow him." What happened next is not recorded but the regiment certainly did not return to the battle line. And it was a dangerous job trying to turn a fleeing mob. When General Hardy tried to rally one fleeing regiment they fired at him. This early incident of "fragging" symbolized the essential lack of discipline in what were basically huge militias, rather than fully trained military units.

The cowardice and "straggling" at Shiloh required immediate investigation. The Southern authorities found that it amounted to a lack of moral fiber on the part of many of their soldiers. The men were not simply exhausted or disillusioned; many of them were quite prepared to admit that they were cowards and did not want to fight. They were clearly the wrong sort of men to place in a battle line. Colonel Strahl wrote, as part of his official report:

> On Monday morning we . . . had a great number of stragglers attached to us. The stragglers demonstrated very clearly this morning that they had strayed from their own regiments because they did not want to fight. My men fought gallantly until the stragglers ran and left them and began firing from the rear over their heads. They were then compelled to fall to the rear. I rallied them several times and . . . finally left out the stragglers.[45]

That doyen of Confederate commanders, P.G.T. Beauregard, joined the debate about Shiloh, when he wrote, "Some officers, non-commissioned officers, and men abandoned their colors early in the first day to pillage the captured encampments; others retired shamefully from the field on both days, while the thunder of cannon and the roar and rattle of musketry told them that their brothers were being slaughtered by the fresh legions of the enemy."

During the siege of Vicksburg in 1863, the Confederate troops behaved with "shameful neglect." Colonel Edward Goodwin reported, "At this time our friends gave way and came rushing to the rear panic-stricken . . . I brought my regiment to the 'charge bayonets,' but even this could not check them in their flight. The colors of three regiments passed through . . . We collared them, begged them, and abused them in vain." During Jubal Early's campaign in the Shenandoah Valley in 1864, Wiley refers us to "some of the most disgraceful running of Confederate history."

[45]B. I. Wiley, *Johnny Reb,* p. 84.

During the engagement at Winchester, Stephen Ramseur recorded the following comment on his own regiment's performance: "My men behaved shamefully—they ran from the enemy . . . the entire command stampeded. I tried in vain to rally them and even after the Yankees were checked by a few men I posted behind a stone wall, they continued to run all the way to the breastworks at Winchester—and many of them threw away their guns and ran on to Newtown six miles beyond. They acted cowardly and I told them so." More trouble followed at Winchester where General Grimes threatened "to blow the brains out of the first man who left ranks." He belabored the cowards with the flat of his sword blade but it seemed to have little effect. Even the ladies of Winchester "came into the streets and begged them crying bitterly to make a stand for their sakes if not for their own honor." These were wasted tears. The cowards "did not have the shame to make a pretence of halting." Indeed, at this stage of the war, fleeing regiments comprised a substantial problem for Southern officers. Morale was sinking and though some Confederate soldiers fought with their accustomed valor, an increasing proportion of Confederate regiments were unwilling to risk their lives in a losing cause. Early seems to have been particularly unfortunate with his men. At Cedar Creek in 1864 he suffered when his men turned a brilliant victory into a defeat by pillaging the enemy camp in search of loot. Even officers joined their men in ransacking the Union stores. While thus engaged Early's men were hit by a Union counterattack and broke in chaotic rout. Of this farcical collapse Grimes wrote, "It was the hardest day's work I ever engaged in, trying to rally the men. [I] took over flags at different times, begging, commanding, entreating the men to rally—[I] would ride up and down the lines, beseeching them by all they held sacred and dear to stop and fight, but without any success."

It would be tedious to continue the story of Confederate "skulking" and panicking; the cases are almost numberless. The surprise must not be that the South lost the Civil War but that they did not lose it sooner.

One of the worst and most decisive examples of mass

hysteria and desertion in battle occurred at Chattanooga in 1863, during the Union assault on Missionary Ridge. After the Confederate victory at Chickamauga on 20 September, 1863, the retreating Union forces of General Rosecrans were besieged in the city of Chattanooga. Bragg's Confederate army held a dominant position on Lookout Mountain and along Missionary Ridge, where their artillery batteries controlled the approaches to the city. It seemed only a matter of time before the city surrendered and the Union Army of the Cumberland was captured. Rosecrans had hardly recovered from his disastrous handling of his troops at Chickamauga and was planning a desperate retreat from Chattanooga that could have turned into a total rout. Instead, General Ulysses Grant sent a telegram to Rosecrans removing him from his command and ordering George Thomas, the "Rock of Chickamauga," to take over. Grant knew his man, and ordered Thomas to hold Chattanooga "at all costs." Far from retreating, Grant intended to reinforce the Union position in Tennessee and use it as a springboard against Bragg's army. Grant must have been delighted at receiving Thomas's reply: "We will hold the town until we starve." In spite of the disaster at Chickamauga it seemed that there was still some fighting spirit in the Army of the Cumberland. With George Thomas to lead them they had something to prove to the new Union commander. Meanwhile, Bragg was doing everything possible to help Grant. Dissension within the Confederate leadership, notably between Bragg and Longstreet—on detached service from Lee's Army of North Virginia—who could hardly forgive Bragg for not pursuing the beaten Union troops after Chickamauga. In spite of the proximity of strong Union forces, Bragg decided to detach Longstreet with 20,000 men to campaign against Burnside in eastern Tennessee. Almost certainly Bragg was looking for an opportunity to get Longstreet out of the way, even at the expense of weakening his army on Missionary Ridge from 60,000 to 40,000. It was a rash decision that was to have serious consequences. At the same time as Bragg was weakening his own command, Grant was moving in reinforcements to support Thomas in Chattanooga. Sherman was advancing from Memphis with 20,000 men and "Fighting Joe" Hooker had brought in 12,000 by rail from Virginia. Within a matter of a few weeks Bragg found himself outnumbered by 61,000 to 40,000. Admittedly, Bragg's men were strongly entrenched along Missionary Ridge, but his men's morale must have fallen from the highpoint of Chickamauga as they realized that while they sat doing nothing the Unionists had recovered their nerve and again outnumbered them.

Grant now gave Thomas the go-ahead to turn defense into attack. At the outset, Thomas drove the Confederate outposts from around Chattanooga back to the foot of the steep slope of Missionary Ridge. In spite of his confidence Grant did not immediately plan to scale the five-hundred-foot slopes and assault the Confederate trenches on the crest. This would have been suicidal in his view, little less than Pickett's famous charge at Cemetery Hill on the third day at Gettysburg. Grant was no romantic; he did not want his obituaries decorated with glorious futility. Instead, Grant planned to turn the flanks of Bragg's army. Hooker would begin by taking his men up Lookout Mountain and clearing the garrison there. In fact, though the climb was a difficult one and in spite of the action earning the romantic title "the battle above the clouds," Bragg had seriously weakened the position when he detached Longstreet, and Hooker simply overran its three thousand defenders by weight of numbers. Before night fell, Lookout Mountain was in Union hands. So far Grant was satisfied. For the next day—25 November—he had ordered Sherman to attack the right of Bragg's army, while Hooker descended Lookout Mountain and attacked the Confederate left. In the meantime, Thomas's Army of the Cumberland would stand fast at the base of Missionary Ridge, until the two flank attacks were fully developed. Grant's idea was that the presence of Thomas's men would pin Bragg to his ridge and prevent him reinforcing his flanks. The last thing that entered his mind was that the Army of the Cumberland would charge up a five-hundred-foot hill against a heavily entrenched foe.

In fact, Grant's psychology let him down. Thomas might have warned him about how the rank and file of his army would react to being left standing while other troops—brought in from outside—fought the men who had whipped them at Chickamauga. The Army of the Cumberland had run under Rosecrans; now they wanted to show that they could fight. If an army can be said to have a "mind," this one did. So many of its men seemed to share the same conviction that the army's action actually took on a unity that has rarely been achieved before on such a scale. It was like the British at Inkermann, but on an even larger scale and against an apparently more determined enemy.

While Grant and Thomas watched from a nearby hill, known as Orchard Knob, twenty thousand men from the Army of the Cumberland marched up the slight slope and overran the first Confederate line of trenches. This was the extent of their target but what followed was quite inexplicable. Admittedly, the Confederate troops had panicked and fled from the first trenches, but no order was given for the Union troops to pursue them. However, amidst the fire of cannon from their own artillery, the men in blue chased the fleeing men in gray up the slopes toward the top. It was like an enormous wave. In some places officers led, in other they followed. It hardly seemed to matter. In the words of the Dupuys:

> This was an army with several scores to settle; with the enemy, with themselves, with the other two armies under Grant's command, and with Grant himself. In their last battle they had been humiliatingly defeated. And now, obviously not trusted by Grant, they had been given a minor role while the men of the Army of Tennessee and of the Army of the Potomac had been assigned the major tasks. For almost two days they had fretted at the foot of that hill, waiting for an order which they thought would never come.
>
> Grant watched this movement with growing alarm. He turned to Thomas and asked who had ordered the attack. Thomas replied that he hadn't. Helpless, cigar gripped firmly between his teeth, Grant watched the unprecedented spectacle of mass psychology.[46]

[46]R. & T. Dupuy, *The Compact History of the Civil War,* p. 261.

George Thomas had given no order but General Gordon Granger guessed what the answer was. As he told Grant, "They started up without orders. When those fellows get started all hell can't stop them." The Confederates tried to stop them at first, pouring a hail of fire into the approaching Union ranks but it was to no avail. As this extract from the pen of Major Connolly shows, there was no stopping the men of the Army of the Cumberland that day.

> One flag bearer, on hands and knees, is seen away in advance of the whole line; he crawls and climbs towards a rebel flag he sees waving above him, he gets within a few feet of it and hides behind a fallen log while he waves his flag defiantly until it almost touches the rebel flag; his regiment follows him as fast as it can; in a few moments another flag bearer gets just as near the summit at another point, and his regiment soon gets to him, but these two regiments dare not go the next twenty feet or they would be annihilated, so they crouch there and are safe from the rebels above them, who would have to rise up, to fire down at them, and so expose themselves to the fire of our fellows who are climbing up the mountain.
>
> The suspense is greater, if possible, than that with which we viewed the storming of Lookout. If we can gain that Ridge; if we can scale those breastworks, the rebel army is routed, everything is lost for them, but if we cannot scale the works few of us will get down this mountain side and back to the shelter of the woods. But a third flag and regiment reaches the other two; all eyes are turned there; the men away above us look like great ants crawling up, crouching on the outside of the rebel breastworks. One of our flags seems to be moving; look! look! look! Up! Up! Up! it goes and is planted on the rebel works; apparently quicker than I can write it the 3 flags and 3 regiments are up, the close fighting is terrific; other flags go up and over at different points along the mountain top—the batteries have ceased, for friend and foe are mixed in a surging mass, in a few moments the flags of 60 Yankee regiments float along Mission Ridge from one end to the other, the enemy are plunging down the Eastern slope of the Ridge and our men are in hot pursuit, but darkness comes too soon and the pursuit must cease; we go back to the summit of the Ridge

> and there behold our trophies—dead and wounded rebels under our feet by hundreds, cannon by scores scattered up and down the Ridge with yelling soldiers astraddle them, rebel flags lying around in profusion, and soldiers and officers completely and frantically drunk with excitement. Four more hours of daylight, after we gained that Ridge, would not have left two pieces of Bragg's army together.[47]

The collapse of the Confederate troops was total and as inexplicable in its way as the charge of the Union forces. It was an example of one side imposing its will totally on its enemy. In terms of mass psychology, the Confederate troops had accepted the moral superiority of their opponents and fled. Officers tried to rally their men as officers always will but to no avail. Bragg later reported that:

> no satisfactory excuse can possibly be given for the shameful conduct of the troops on our left in allowing their line to be penetrated. The position was one which ought to have been held by a line of skirmishers against any assaulting column, and wherever resistance was made the enemy fled in disorder after suffering heavy loss. Those who reached the ridge did so in a condition of exhaustion from the great physical exertion in climbing which rendered them powerless, and the slightest effort would have destroyed them.[48]

As the above example shows the moral of the Confederate forces became increasingly brittle as the Civil War progressed. Most of the Southern soldiers had not realized what "total war" would be like. And Ulysses Grant and Abraham Lincoln made an awesome pair of opponents, in that both understood the demands of a modern war and the sacrifices that would have to be made. Neither expected easy victories or a swift solution. Too many Confederate soldiers expected the North's military inexperience to cost them dear—which indeed it did during 1862. But in an

---

[47] H. S. Commager, *The Blue and the Gray, Volume 2,* pp. 911–12.
[48] *Battles and Leaders of the Civil War, Volume 3,* p. 727.

industrial age decisive victories were elusive. What cracked at Chattanooga was not simply the morale of Bragg's army but the morale of the Southern soldier. From this time onward there is a fatalism about so many Southerners. For the first time they have come to realize that their early victories over men like Pope, Burnside, and Hooker have settled nothing. As the power of the Confederacy drained away in costly victories, won by the futile courage of their brave infantry, the strength of the North was only beginning to make itself apparent, in the shape of economic strangulation of the South. For all the romantic and chivalrous achievements of men like "JEB" Stuart and Robert E. Lee, the industrial power of the North was irresistible in the long run.

If the strength of the North was in the ledger books of its actuaries, the strength of the South was in the minds of its people. And once the Southerner ceased to believe in his chances of winning the war his fighting spirit evaporated very quickly. Whereas Bell Wiley demonstrates the overweening confidence the Southern soldiers possessed in the early part of the war, as time went by without total victory they became more realistic—and with realism came despair. Early enthusiasm for the war brought forth a flood of volunteers, expecting a quick victory over the Northerners who, it was presumed, could not fight. But once it became apparent that a triumphant march on Washington was not imminent, many Southerners lost interest in the fighting. Recruitment became difficult and when conscription was eventually introduced in April 1862, many men found ways to avoid the draft. One of the most popular ways was for their physicians to issue certificates of disability to perfectly fit individuals. As one Southerner wrote, conscription would "do away with all the patriotism we have. Whenever men are forced to fight they take no personal interest in it . . . My private opinion is that our Confederacy is gone up, or will go up soon . . . a more oppressive law was never enacted in the most uncivilized country or by the worst of despots." But the law was not strictly applied and men were allowed to purchase substitutes to take their place in the army, resulting in the use of poor quality material. The substitutes—some of whom were aliens who had no interest in the success of the Confederacy—were the first to run in battle and the first to desert when an opportunity arose.

After the battle of Gettysburg which for many Southerners marked the turn of the tide in the war, the realization among the frontline troops that there were large numbers of able-bodied men at home in the South who, by deferment or substitution, were avoiding service in the war was highly destructive of morale. The poorer classes, notably, began to see it as "a rich man's war, but a poor man's fight" and this resulted in divisions in the Confederate armies on the grounds of class. The decline in morale at the front line was reflected in the letters that were sent home to loved ones. War weariness became common in the latter part of 1863. The wild claims that the Yankees would be easily whipped were by now revealed for the nonsense that they so obviously were. Moreover, the lack of food and clothes on the part of the soldiers was debilitating. Many Confederate soldiers marched barefoot. This has been too often seen as an example of their toughness and staying power rather than as an obvious sign of the breakdown of the Southern logistics system. Lack of supplies became a prevalent excuse for desertion on a mass scale.

An alarming sign of the loss of morale was the incidence of self-inflicted wounds in the Confederate armies. During one skirmish it was reported that a Confederate soldier hid behind a tree and waved his arms up and down on either side. When asked by an officer what he thought he was doing, he replied that he was "feeling for a furlough," meaning that he was hoping for a slight wound that would enable him to convalesce away from the front. By 1864 war weariness in Lee's army was widespread. As one man wrote, "The men is not a'going to stay on the field any longer for they say that they have fought long enough and that they will not fight any more."

Poor food was particularly destructive of morale. Early in the war, the Rebel soldiers had boasted that they could march and fight on an empty stomach. But as defeats accumulated and lesser men filled the places of the men killed, the absence of proper rations was unacceptable. As one private soldier wrote home in 1862, "If I ever lose my patriotism and the 'Secess' spirit dies out then you may know the Commissary is at fault." Subsistence on cornmeal mixed with water, and tough beef three times a day, was hardly enough to sustain the hard-marching, hard-fighting Johnny Reb for very long. In fact, this diet did more to undermine the Confederate soldier than any amount of Yankee bullets. Significantly, by 1863 some of Bragg's troops were already refusing to obey orders on account of their poor rations. An even more damaging blow to the soldiers' morale was the knowledge that their families at home were suffering the same privations as they were. One Alabamian wrote to his wife in October 1863 that he would try to get a furlough but if he could not he would come home anyway as he could not bear to hear that she and her children were "suffering for bread." Temptations to desert were growing all the time. It was difficult for officers to issue furloughs as there was little chance of men returning across the Mississippi once they had left the camp. One officer even went as far as crossing the river and rounding up four hundred deserters, but when he tried to return with them all but twenty-five escaped.

Fundamentally, Southerners—for all their undoubted fighting qualities—did not make good soldiers. Their cantankerous individualism made them unwilling to accept the authority of their officers. One Texan demonstrated this in a letter home to his folks:

> We see hard times, as you can guess it is tolerable bad for me as I am not allowed the chance of a dog. I have come to the conclusion that I will stay and tuff it out with Col. Young and then he can go to Hell for my part. You know that if anyone will try to do I can get along with them but when they get Hell in their neck I can't do anything with them and so I don't try. If a man treats me well I will stick up to him till I die and then see that my spirit helps him when I am gone to my long home but he has acted the damned dog and I can't tell him so. If I do they will put me in the Guard House . . . but I can tell him what I think

> of him when this war ends and as to go with him I won't do it to save his life . . . I will come [home] when my time is out or die. I won't be run over no longer not to please no officers. They have acted the rascal with me . . . I am so sick of war that I don't want to hear it any more till old Abe's time is out and then let a man say war to me and I'll choke him.[49]

The Army of the Confederacy was broken in spirit long before the final surrender at Appomattox. After their defeat at Gettysburg the roads of Virginia were crowded with deserters. When asked to show their furloughs, some of them simply tapped their rifles and replied that these were the only furloughs they needed. Desertion was so prevalent that it is estimated that as the war drew to a close nearly two-thirds of the entire Confederate Army was absent from the ranks. By February 1865, at least a hundred thousand men were currently listed as deserters. Shortly afterward official statistics showed Confederate armies with 160,198 men present for duty and a staggering 198,494 absent for one reason or another. Johnny Reb had voted with his feet. He had expected the war to be easy. It had not turned out that way. The despised Northerners—Billy Yank to him—had shown more fighting spirit than he had expected. In fact, Billy Yank had proved himself to be a good soldier, Johnny Reb—as he showed at Missionary Ridge—was a great talker, a great fighter, a great marcher but—sometimes—a great runner when the going got too tough.

To set the record straight, Billy Yank did some running too and, one might think, with rather less reason on occasions. Even with the war going entirely in his favor Billy Yank sometimes decided to sit it out and let other men do the fighting. Bell Wiley's trendsetting account of the Unionist soldier contains as much evidence of Northern cowardice and desertion as it does of Confederate shirking. It would be misrepresenting the men of the Northern states if one left the impression that the charge by the Army of the Cumberland was anything other than what it was—an

[49]B. I. Wiley, *Johnny Reb,* p. 140.

unparalleled mass action in which, at least temporarily, fighting spirit reached amazing heights. "Unparalleled"—it should be noted—as applied to the American Civil War. Pickett's Charge at Gettsburg seems to me, as I explained above, a more straightforward brand of heroics based on men nobly following orders. In that way Pickett's Charge is more reminiscent of the "Charge of the Light Brigade," whereas the Charge of the Army of the Cumberland resembles more the action of the British infantry at Minden.

Enough of heroics; they paint a small part of the true picture. The Civil War was a period of initial enthusiasm waning as the realities of modern war were brought home to a basically civilian army on each side. We have seen how the Confederates reacted to the stresses of war. Did the Unionists behave any better? In a word: No.

Both sides were new to war at First Bull Run and so perhaps both can be excused the mass stampede to the rear that occurred on each side. However, by the battle of Shiloh in 1862, one might have expected that discipline would have asserted itself. Unfortunately, one would have been mistaken, for Grant's army abandoned their posts "in their thousands." One entire Ohio regiment ran, "in a manner that can only be stigmatized as disgraceful and cowardly . . . their officers setting them an example of speed in flying." At the same battle General William Nelson related, "I found cowering under the river bank when I crossed from 7,000 to 10,000 men frantic with fright and utterly demoralized, who received my gallant division with cries of 'We are whipped; cut to pieces.' They were insensible to shame or sarcasm." Hundreds of Union troops, and officers, were seen crossing the river on logs, to escape the fighting. After the battle of Fredericksburg, in the same year, Lieutenant Henry Ropes reported that hundreds of Hooker's division "ran by us like Sheep. I saw a whole brigade of Pennsylvania cowards (Tyler's Brigade) break and run in total disorder when they were brought up to our relief, our men cursing them most heartily." General Rosecrans accused the 17th Iowa Regiment of "disgraceful stampeding" at Iuka, while at the battle of Perryville, a Union officer reported that General Jackson's troops "turned and fled at the first fire." The major blame was placed on the officers, of whom even colonels were seen panicking and running. One regiment panicked so far as to fire a volley into its neighboring regiment before breaking and fleeing. At the battle of Murfreesboro, General McCook's troops panicked, while Van Cleeve's division fled from the field. An observer wrote, "It was difficult to say which was running away the more rapidly, the division of Van Cleeve to the rear or the enemy in the opposite direction." Perhaps the worst example of Billy Yank running away was in 1863 at the battle of Chancellorsville, as a consequence of "Stonewall" Jackson's surprise attack on the Union rear. The XIth Corps was guilty—almost to a man—of "arrant cowardice." At Sabine Cross Roads, the vanguard of the Union Army—some 8,000 strong—suddenly panicked and fell back on the units behind them yelling, "All is lost!" No reason for this action can be suggested: mass hysteria as usual is probably the explanation. During the Wilderness Campaign of 1864, Grant's army experienced some of the largest desertions of the entire Civil War period. Thousands of his troops took refuge in the forests rather than engage the enemy. Low morale as a result of the appalling fighting conditions help to explain the high level of desertions. Even the regular use of executions for deserters was not enough to deter the determined skulker. Billy Yank was no more willing a conscript than Johnny Reb. Once battle fatigue set in, it did not take much to overturn the very limited confidence of many Union units. Although higher morale might have been expected of the Union forces after 1863, battle exhaustion was a very real threat to combat performance right until the end of the fighting.

8

"Pals" on the Somme—1 July, 1916

S. L. A. Marshall has conclusively shown that the strongest prop for the fighting soldier is the support of his friends and comrades. This truth has been known—if not so clearly expounded—for generations and, perhaps, the bravest and most tragic manifestation of it was the setting up of the "Pals battalions" in Britain during the transformation of the Kitchener volunteers of 1914 into the British Army that fought on the Somme two years later.

The flood of volunteers to fight for "King and Country" in response to Field Marshal Kitchener's appeal in 1914 was, in the words of historian John Keegan, "a spontaneous and genuinely popular mass movement which has no counterpart in the modern English-speaking world and perhaps could have none outside its own time and place: a time of intense, almost mystical patriotism and of the inarticulate élitism of an imperial power's working class . . ." So many men volunteered that the recruiting system was unable to cope. Nevertheless, the people, along with Kitchener alone of British politicians and generals, seemed to realize the immense struggle that was about to ensue. A modern war between industrial nations would not end quickly. Upper-class Britons might be joining up so as not to "miss the fun" and be promising their loved ones that "it'll all be over by Christmas," but Kitchener was preparing for a long, attritional war, in which the masses would provide the nation's military sinew rather than a few Guardsmen with lances, fresh from chasing foxes, and trying to act out their dreams of glory. It was from the industrial cities of Northern Britain that the nation's lifeblood came and Kitchener responded to the enthusiasm of Northern workingmen by allowing volunteers to choose their own titles for their units, join up with their friends and colleagues, and serve together in the same battalion. This was the beginning of the "Pals battalions"—an inspired idea, if there ever was one—but one that ended in tragedy.

The "Pals" came from every walk of life, but the vast majority were working class in origin. They came from "work-places, factories, unions, churches, chapels, charitable organizations, benefit clubs, Boy Scouts, Boy's Brigades, Sunday Schools, cricket, football, rugby, skittle clubs, old boys' societies, city offices, municipal departments, craft guilds—any one of those hundreds of bodies from which the Edwardian Briton drew his security and sense of identity." Faced with this kind of response, Kitchener guaranteed that those "who joined together should serve together."

The idea that had originated in the North of England caught on elsewhere. Soon "Pals" units were blossoming throughout the country. What an army they promised to make, built with the bricks of friendship and group identity. From Welsh holiday resorts to London slum boroughs, men poured forth in their battalions with the unlikely names of the First Public Works Battalion, the Forest of Dean Pioneers, the Civil Service Battalion, the Grimsby Chums, the North-East Railway Battalion, the Artists and Writers' Battalion; there seemed no end to the ingenuity of the British workingman and his willingness to serve his King and Country. Even "Bantam" units were set up for the miners, many of whom, though tough as nails, were very short and much below the minimum height allowed for normal regiments. But the enthusiasm of the volunteers soon overran the capacity of the authorities to cope with them all. Kitchener had thought it quite enough to get men to volunteer. After that, their training was somebody else's problem. And a large problem it proved to be. As John Keegan has shown, the name "battalion" was often a very loose description of these military units. Sometimes they consisted of no more than about a thousand men who arrived together by train. Their military skills were minimal and their uniforms and equipment nonexistent. What was even more of a problem was how did one find enough experienced officers for these thronging multitudes? At first the Indian Army was combed for suitable men—500 happened to be on leave in Britain in 1914 and they were all commandeered for new regiments—but once these men had been taken it was necessary sometimes to call up retired pensioners or even disabled men to put the new recruits through their paces. One of the most tragic methods of selecting new officers was that of preparing a list of 2,000 "young gentlemen" who had just left the best of England's schools or universities. These young men were immediately offered a commission simply on the assumption that men from their background—educationally, socially, and financially—would inevitably be "officer-material." Courage these youngsters had in buckets but of common sense or even natural caution they had little. The great schools of England—Eton, Winchester, Harrow, Westminster, Rugby, Charterhouse, Marlborough, Wellington and their like—all contributed their sixth form leavers willingly. And each of these schools and hundreds like them today boast their well-oiled plaques recording those who died in the Great War "For King and Country."

To the recruits, however, with their dummy rifles and blue serge uniforms, it all seemed like a great adventure. Without accommodation in barracks, many of the "Pals" had to live at home. Each day they would assemble in a field or on a hillside and train to be soldiers, dressed in their civilian clothes and carrying broomsticks for guns. They were "Kitchener's Army" and they were intensely proud. The spirit of the "Pals" can be seen in this little poem written by one of the Bradford "Pals" and published in the *Bradford Daily Telegraph,* just days before news of the tragic losses of 1 July, 1916, reached Britain:

We gets our rum and lime juice
We gets our bully beef,
Half a dozen biscuits
That break your bally teeth.
We get no eggs for breakfast,

But they send us over shells,
And you dive into your dug-out
And get laughed at by your pals . . .[50]

Local pride in their own "Pals" was enormous and as Private George Morgan of the Bradford "Pals" wrote the fighting spirit of the "Pals" was remarkable.

> The companionship was marvellous, absolutely marvellous. Everyone seemed to help one another and agree with one another. It was lovely. We were all pals, we were happy, very happy together: and they were such good people. They were fine young men, the cream of the country. That spirit lasted until 1 July, 1916. We had so many casualties that we were all strangers after that. The new men who came were fed up, they were conscripts and they didn't want to come, they didn't want to fight. Things were never the same any more.[51]

But in a sense the happiness was born of ignorance. In historian John Keegan's words:

> The promise of tragedy loomed about these bands of uniformed innocents [and] was further heightened by reason of their narrowly territorial recruitment; what had been a consolation for the pangs of parting from home—that they were all Pals or Chums together from the same close network of little city terraces or steep-stacked row of miners' cottages—threatened home with a catastrophe of heartbreak the closer they neared a real encounter with the enemy.[52]

None of these volunteers saw action until 1 July, and after that day nothing could ever be the same again. The massive losses suffered by some "Pals" units—many suffered fifty percent casualties—meant that they had to be made up to strength by feeding in replacements from any

---

[50]M. Brown, *Tommy Goes to War,* p. 36.
[51]Ibid., p. 194.
[52]J. Keegan, *The Face of Battle,* p. 226.

source available. The "Pals" idea collapsed and so bitter were the memories for the survivors of losing so many good friends at one time that the fighting spirit of the units could hardly be restored.

The first of July, 1916, was the blackest day in the history of the British army. It was a day that ended forever the idea that British infantry possessed the moral advantage over their enemies that they believed had been theirs by right since the days of Fontenoy and Minden in the eighteenth century. The Boer War had severely dented this belief but after the BEF's heroic displays at Mons and First Ypres, there were those who believed that the sight of British troops in line abreast marching toward them was still enough to put any enemy to flight. The Germans, never subscribers to this whimsical British conviction after witnessing their antics in South Africa in 1900, shrugged their shoulders and mowed them down in their thousands. They were the professionals. The British, clearly the "amateurs." And among the sixty thousand casualties the British incurred on 1 July, many who fell were from the "Pals" battalions, enthusiastic but half-trained civilians, fighting and dying with their friends for company. The architect of the "Pals" was spared the sight of his courageous horde being squandered on the battlefield, before most of them had even fired a shot in anger. Lord Kitchener was drowned when the ship in which he was traveling to Russia—H.M.S. *Hampshire*—struck a German mine and sank. Instead, the architects of 1 July—for there were two—were Sir Douglas Haig, the British commander-in-chief and Sir Henry Rawlinson, commander of the British Fourth Army. On 1 July, 1916, Rawlinson, in fact, was commanding probably the biggest army—the Fourth Army had over 500,000 men—that any British general had ever commanded. The great responsibility, unfortunately, seemed to reveal cracks in his own moral courage. He had been given a priceless gift by his country—a huge army of British volunteers—with which to launch what was hoped would be the war-winning offensive against the Germans. But was Rawlinson—or indeed Haig—man enough to use it?

If one wishes to understand the extraordinarily high morale of the British army that "went over the top" on 1 July, 1916, it is necessary not only to understand the mentality of the men who fought but the nature of the war as the British public saw it. The British media was a far more powerful tool in the hands of politicians than was the case in any other of the warring nations. Its message, even in difficult times, was clear. All was well and Britain was winning the war. The newspapers spoke of Britain's soldiers singing as they marched into battle; looking forward to "a scrap" with the Huns, and inspired by the noblest of patriotic thoughts. And these lies—for lies they were as the soldiers at the front well knew—were fed to the general public at home so effectively that the British people were almost certainly the most belligerent of all the combatants. Soldiers returning home on leave found the civilians they met bristling with hatred of the Germans, a hatred that the frontline soldiers never shared. And the Kitchener "volunteers" training in Britain for action in 1916 could hardly fail to absorb the propaganda that was directed at them by the media. As a result, the men of 1916 had none of the cynicism of the French conscripts who had been fighting since August 1914, nor of the British regulars who had all been wiped out by Christmas 1914. They had not yet been "blooded" and they possessed enthusiasm for war that only a civilian could feel. It took just a single day for this naïveté to die, but the cost of exorcising it forever was truly terrible in dead and injured.

Yet if Haig and his generals were butchers by profession, it was through necessity rather than from a lack of imagination. The strategic situation required a "grand gesture" by Britain, if only to demonstrate to her allies that she was prepared to bear her share of the common burden. By the end of 1915 it was apparent to the British government that there was no prospect of the war against the Central Powers being won without Britain assuming a much greater share of the fighting. French losses in their spring and autumn offensives of 1915 had stretched their manpower to its limits. It was obvious that the vast numbers of Britons who had responded to Lord Kitchener's appeal for volunteers in 1914 would now need to be used in France. British politicians did not reach this conclusion without a struggle. During 1915 attempts had been made to find an alternative winning solution by undertaking campaigns in Gallipoli and Salonika. But in neither of these peripheral areas was Britain able to force the Germans to withdraw troops from the main battleground in Belgium and France. Thus when the Allies met at Chantilly to coordinate their strategy for 1916 Britain was obliged to commit herself to a summer offensive in France, in conjunction with a major French offensive there.

On 7 April, 1916, the British government authorized their new commander-in-chief, Sir Douglas Haig, to concert an offensive with the French. The battlefield was to be the Somme region—in the words of Sir Henry Rawlinson, "The country resembled Salisbury Plain, with large open rolling features, and any number of partridges which we are not allowed to shoot." In fact, apart from the partridge-shooting, the Somme was a very bad choice. The German front lines there enjoyed the advantage of the high ground, so that the waves of British soldiers advancing would face a climb toward the strategically vital Pozières ridge. In addition, the chalky subsoil had allowed the Germans—unchallenged in the area since October 1914—to construct intricate and effective underground defenses, resistant to artillery barrage. Nevertheless, the Somme was the area chosen by French commander Marshal Joseph Joffre, and Haig felt obliged to comply. At first, Joffre had planned an operation in which equal numbers of French and British troops would be involved, but after 21 February—when the Germans began their massive assault on Verdun—the French commitment to the Somme diminished. By May, Joffre was only able to offer thirteen divisions to support the twenty or so British divisions earmarked for the assault. By the late spring of 1916 British strategic options had narrowed. With the pressure around Verdun threatening to break the French army Haig had little alternative but to attempt to lift the siege by diverting German troops to the Somme.

But what sort of offensive were the British planning? Experience at Loos and Neuve Chapelle in 1915 had demonstrated that unsupported infantry and cavalry had no chance in the contested zone while enemy firepower remained unsuppressed. Moreover, territorial gains were only temporary if the Germans were able to feed in reserves at threatened points of their line and to stage effective counterattacks. The only solution seemed to rest in firepower. On the Somme it would be necessary to overwhelm the German defenses with an artillery bombardment of unprecedented weight and ferocity. Once the artillery had succeeded in demolishing the enemy it would be a relatively simple task for the infantry to occupy the ground won. It all sounded very simple.

Haig had chosen Fourth Army commander Sir Henry Rawlinson to spearhead the offensive on the Somme. With the French on their right, Rawlinson's army was to attack along an eighteen-mile front between Gommecourt in the north and Montauban in the south. Half a million men under his command would prepare the most prodigious military operation ever undertaken by a British army. In some ways Rawlinson—an efficient administrator and an effective infantry general—was a good choice, yet in one vital element he did not see eye to eye with Haig about the forthcoming battle. And in this disagreement between commander-in-chief and army commander lay the seed of a disaster.

A staggering amount of preparation was needed behind the British lines in the weeks before the operation was to take place. A whole new town had to be built to accommodate the assault troops, to feed and equip them, to provide them with medical services and entertainment. Hundreds of thousands of horses and transport vehicles passed endlessly up the newly constructed roads and railways, and field guns, howitzers, and mortars in unprecedented quantities were moved into position and hidden from prying German eyes. Yet how could all this preparation take place in secrecy? In the first place, the Royal Flying Corps drafted planes into the Somme region and established air supremacy over the British lines to keep German reconnaissance aircraft at a distance. In France every effort was made to keep the assault secret. But in Britain no such secrecy prevailed. All efforts to keep the Germans guessing were ruined when British newspapers reported that munitions workers had had their Whitsun leave canceled and been put on round-the-clock rotas. Nothing could be clearer to the Germans than that the British were planning a major offensive. But where would it come? In Flanders, where the bulk of the British troops were based, or farther south? It did not require a genius to conclude that British aerial activity on the Somme was an attempt to conceal major troop concentrations there. As a result, the Germans were quite prepared for the British attack whenever it might come. They concluded that they could rely on the British generals to give them ample warning. In any event, Rawlinson's seven-day artillery barrage was as clear a signal as anyone might need.

Infantry training was taken very seriously by the officers but probably raised more than a few laughs from the Tommies with a feel for black comedy. Behind the lines a huge area of sandy, dusty soil was laid out with tapes. The men then went "over the top" and captured the "tapes"—it was all bloodless and very symbolic. As the Tommies went into action they had to imagine gas and clamber over imaginary barbed wire. To make matters worse everyone had to walk as if he were fully loaded with seventy pounds of kit, wearing gas masks, carrying spades, pigeons, rolls of barbed wire, and extra bombs. Such training was considered necessary by the officers, who felt that the men could not be trusted to do anything as simple as run in a straight line or take cover in shell holes.

While the soldiers toiled, their commanders were at odds with each other as to what exactly was the aim of the offensive. Haig was still thinking in terms of a "decisive battle," which would break the German lines and enable him to use his massed cavalry corps to burst through into open ground. Thus Haig hoped to punch a hole and pour through it in a grand Napoleonic sweep. Unfortunately for him, and perhaps for his men, Rawlinson did not agree with him. He had lost hope of a breakthrough and preferred to think in terms of a "bite and hold" operation. He rejected Haig's belief in "going for the big thing" and, instead, aimed to deliver sharp blows against limited targets, that would exhaust the German reserves as they tried to repair holes at various points in their line. So, in spite of Haig's own belief that the Fourth Army was aiming for a decisive breakthrough of both the first and second German defensive lines on 1 July, it was in fact merely proposing to take the first positions, before consolidating and moving up the artillery to support further advances. How Haig and Rawlinson could have arrived at such different interpretations of their task after such lengthy discussions is difficult to understand.

As alarming—and even more disastrous in its human consequences—was Rawlinson's application of infantry tactics. As a professional soldier he seemed to have little faith in the men who would compose the assault force on 1 July. Sixty percent of them would be men from Kitchener's "New Army"—volunteers and ex-civilians—who in his eyes could not be trusted to do anything properly. As a result, whereas the French infantry on the British right advanced across no-man's-land in small, tight groups, rushing from one piece of dead ground to another, covered all the time by other groups behind them, the British soldiers—weighed down by seventy pounds of equipment—were ordered to advance upright, a yard or two apart from their neighbors and at a walking pace, to prevent them panicking and diving for cover. A training memorandum issued just three weeks before the attack by Haig's chief of staff, Sir Lancelot Kiggell, ordered the attacking infantry to advance in four rows. Kiggell warned that heavy casualties could be expected and did what he could to guarantee them by having the British advance like little tin soldiers waiting to be swept away by a bored child's arm. The German defenders were later to write in astonishment about the slow, steady march of the British who, had they come at a rush, would certainly have succeeded in capturing many more trenches. Paul Scheyt wrote:

> The English came walking, as though they were going to the theatre or as though they were on a parade ground. We felt they were mad. Our orders were given in complete

> calm and every man took careful aim to avoid wasting ammunition.[53]

Musketier Karl Blenk agreed:

> When the English started advancing we were very worried; they looked as though they must overrun our trenches. We were very surprised to see them walking, we had never seen that before. I could see them everywhere: there were hundreds. The officers were in front. I noticed one of them walking calmly, carrying a walking stick. When we started firing, we just had to load and reload. They went down in their hundreds. You didn't have to aim, we just fired into them. If only they had run they would have overwhelmed us.[54]

A French artillery observer commented, "I thought of the Crimea today, and of what the French said in the Crimea about the Charge of the Light Brigade."

Rawlinson's tactics were a formula for disaster and one might, at this stage, be wondering how such absurd instructions were ever given to soldiers marching in some places as far as a thousand yards into barbed wire, all the time under raking fire from machine guns. In his own defense, Rawlinson would have insisted that this scenario was wrong for the simple reason that there would not be any machine guns, or barbed wire, or even live Germans. The British infantry was advancing to occupy ground won for it by the devastating weight of the British bombardment. And in this assumption lay the core of the disaster that was to strike Fourth Army on 1 July, 1916.

Rawlinson's confidence in the power and effectiveness of his artillery was absolute. As he told his officers, "Nothing could exist at the conclusion of the bombardment in the area covered by it." He was using 1,437 guns on a fifteen-mile front and his guns would achieve three things: they would suppress German artillery fire when the assault

---

[53]M. Middlebrook, *The First Day of the Somme,* p. 157.
[54]Ibid., p. 156.

began; they would destroy the German barbed wire—even though some of it was so thick that light could scarcely pass through its close-meshed coils; it would kill all the German soldiers in their trenches, dugouts and bunkers, so that there would be no one to scythe down the walking waves of Fourth Army. But just how effective would the British bombardment be? Previous barrages had only ever succeeded where the Germans were taken by surprise or had poor bunkers in unsuitable terrain. Was this the case on the Somme? The answer is no. In spite of every effort to keep them in the dark about British intentions, the Germans knew where and when the British attack would take place. Even worse, the Germans had been in their Somme positions for two years and had used the time to build the best and deepest defensive positions on the whole of the Western Front. In places the British would meet four separate trench systems and might have to cross twelve trenches before reaching open country. Concrete dugouts thirty feet deep kept the soldiers in safety during the barrage, and barbed wire entanglements of awesome efficiency lined the forward trenches. This was what Rawlinson was dismissing so airily when he spoke of the irresistible power of his big guns. And his confidence in the guns was quite misplaced. Of the 1,437 artillery pieces available, only 467 were heavy guns and of those just 34 were of 9.2 inch caliber or more. In the event, just thirty tons of explosive were to fall on each mile of the German front—hardly impressive when the distinguished military historian John Keegan has suggested that such powerful defenses would today warrant several small nuclear warheads. Even worse was the nature of the shells that would be fired. Nearly two-thirds would be shrapnel, deadly to men in the open, but harmless to those in deep dugouts. The shrapnel could cut the barbed wire if fired with precision but there is no evidence that the British gunners had developed such skills at this stage. Of the 12,000 tons of explosive fired by British guns in the last week of June, 1916, just 900 tons were of high explosive capable of destroying the deepest German defenses. And when one considers that nearly one third of the shells fired—many of low quality American manufacture—failed to explode at all, one can see how disappointing was this supposedly all-destructive barrage. Nor was accuracy a virtue of the British artillery at this stage of the war. British shells could never be used less than 300 yards ahead of the British troops—but the frontline trenches were often not even that far apart. As a result, much of the artillery fire missed the German front lines, machine gun positions, and concrete pillboxes. The British figure of a 300-yard safety limit should be contrasted with a French one of 60 yards and a Japanese one (in the Russo-Japanese War eleven years before) of 100 yards to illustrate how mistaken Rawlinson and Haig were to rely on the British gunners to provide a "creeping barrage" for their infantry, or to silence the German front lines and their wire defenses. In simple terms, the gunners were not up to the job. Thousands of British troops were to pay for this with their lives.

Now what do you do if you are a commander who has staked everything on the success of an artillery bombardment that has clearly failed? Do you show the normal courage necessary to call off the attack and go back to the drawing board? Or do you simply pretend that everything had happened as you predicted? Both Haig and Rawlinson are culpable in that, in spite of intelligence reports that spoke of unbroken barbed wire facing three of the five corps due to attack on 1 July, and of prisoners taken who had clearly suffered little from the intensive barrage, they went on blithely believing what they wanted to believe rather than the evidence of their own eyes. Haig even remarked fatuously, "The barbed wire has never been so well cut, nor the artillery preparations so thorough." Haig clearly had no real idea of what was going on at the front. On the other hand, Rawlinson himself wrote, "I am not quite satisfied that all the wire has been thoroughly well cut and in places the front trench is not knocked about as I should like to see in the photos. The bit in front of the 34th Division has been rather let off." In fact, it was the 34th Division that was to be massacred on 1 July. With the evidence to hand, Rawlinson could have prevented this, but he did not, ordering the attack to proceed as if the barrage had been totally successful.

The men were told deliberate lies to maintain morale. The Sherwood Foresters were assured: "You will meet nothing but dead and wounded Germans. You will advance on Mouquet Farm and be there by 11 A.M. The field kitchens will follow you and give you a good meal." The King's Own Yorkshire Light Infantry (who were to suffer seventy-six percent casualties) were told: "When you go over the top, you can slope arms, light up your pipes and cigarettes, and march all the way to Pozières before meeting any live Germans."

Among the "Pals," the first signs of disquiet were beginning to appear and some of the camaraderie was breaking down. Private Bloor of the Accrington "Pals" described one unfortunate incident:

> Rumours came back from the front line that the German wire was not all cut and many of our chaps were a bit jumpy. While we were sharpening our bayonets an argument broke out and one of my pals, the sanitary Corporal who was being left behind, joined in. Someone said, "You shut up! You don't have to go in." He felt so bad that he went to his company commander and got permission to come in with us. He was killed.[55]

Here was an example of camaraderie working in reverse. The sanitary corporal had no need to join the assault but felt that he would lose face with his friends and therefore asked permission to go. It cost him his life but had he not gone with them it would have been a serious blow to his self-esteem. It took a sort of courage to join the others; it would have required greater courage for him to have remained behind at his post.

As zero hour approached much of the sham-courage that had sustained the men in front of their peers began to wear thin. Private Dunn of the Durham "Pals" seemed to sense some ulterior motive in having his pay made up. Like many soldiers, he found a sort of relief in the "attractions" of the French *estaminets.*

[55]M. Middlebrook, *The First Day of the Somme,* p. 103.

> Our bank books were taken in and we were paid all our credit, which was substantial. Some of us had hundreds of francs, which immediately gave the impression to many of us that most were not coming back. The result was that the *estaminets* got most of it.[56]

Self-inflicted wounds were more common among experienced soldiers than the well-motivated new arrivals. It was a sign of the stress that everyone was feeling as they awaited zero hour that some men chose this way out of their predicament. Private Seneschall of the Cambridge Battalion was still able to wonder about the fate of individuals at a time when a hundred thousand men were to risk all in the furnace of modern war.

> A shot went off some yards away; a fellow had shot himself right through the knee. He had pluck, I think. It was a strange sight to see him being carried away on a stretcher under arrest, with a man at each side of him with fixed bayonets. I often wonder what happened to him.[57]

As zero hour drew near and everyone marched to his position, Private Slater of the Bradford "Pals" found great consolation in the company of an older man.

> The feeling of comradeship seemed to grow among us as we marched forward into a common danger. In particular I have a lasting memory of the man who was closest to me as we marched. I was only eighteen at the time, having joined the army under age, and he was some years older than I. As he spoke to me I became aware of a feeling almost of tenderness in him towards me, as though he sensed my fears and was trying to reassure both himself and me. "Don't worry, Bill," he said. "We'll be all right." And

[56]M. Middlebrook, *The First Day of the Somme,* p. 103.
[57]Ibid., p. 103.

he spoke as gently as a mother trying to soothe a frightened child.[58]

Rum was issued in the tea—or taken neat—and soon the world seemed a simpler place. Problems that had seemed insurmountable came into perspective, as Private Brownbridge of the Northumberland Fusiliers found.

> I found that I wasn't alone, as a second-lieutenant was standing beside me, shaking like a jelly, which nearly made me jittery myself. He was just a youngster, about my own age, and had just joined the battalion a few days before. I shouted at him to get over the top but he just looked at me forlornly and couldn't seem able to speak. I whipped out my bottle of rum, I had been saving it for several days, and offered it to him but he must have been a teetotaller as he only took a sip. I told him to take a good drink, which he did. You never saw a man find his courage so quickly. He pulled out his revolver, climbed the ladder and went charging after the men like a hare. If we hadn't had our rum, we would have lost the war.[59]

Whistles blew all along the line and from the trenches rose up a legion of men inspired by a single thought: they must leave their trenches and go forward. After that—nothing. They had been told that all the Germans were dead. The artillery had destroyed everything. There would be no machine gun fire, no artillery shells. Then were their ears deceiving them? Why were men falling all around? Had their officers lied? Private Houston of the Public Schools Battalion soon discovered the truth:

> Imagine stumbling over a ploughed field in a thunderstorm, the incessant roar of the guns and flashes as the shells exploded. Multiply all this and you have some idea of the Hell into which we were heading. To me it seemed a hundred times worse than any storm. On top of all this, we were losing a lot of men. When I say men, I should

---

[58] M. Middlebrook, *The First Day of the Somme,* p. 108.
[59] Ibid., p. 160.

> really say boys, because we had been drafted to a battalion of public school boys. They were a nice lot of lads and I hated to think of them going up against trained men of the German Imperial army.[60]

Like the eternal question of whether there is life after death, every soldier wondered if his courage would fail at the vital moment. Only in battle would he know, but would he live to benefit from the knowledge? For many of the young, inexperienced men of Kitchener's "New Army" the answer came that morning of 1 July, 1916. As one ordinary soldier said, "I hadn't gone ten yards before I felt a load fall from me . . . I had been worried by the thought: 'Suppose one should lose one's head and get other men cut up! Suppose one's legs should take fright and refuse to move!' Now I knew it was all right. I shouldn't be frightened and I shouldn't lose my head . . ."

At 7:30 A.M. on 1 July the first waves of 60,000 men went over the top and marched slowly toward the German lines. Some were led by officers who kicked footballs, others by men with walking sticks or umbrellas. It was a jaunt, they had been told. Within thirty minutes half of them had become casualties. Of the 120,000 from 143 battalions who attacked that day, nearly 60,000 casualties were suffered, including some 20,000 dead. It was the greatest loss ever suffered by the British Army and the heaviest by any army in a single day of the entire war. In fact, British battle casualties that day exceeded those from the entire Crimean War, Boer War and Korean War put together. The cause was simple. The German soldiers had survived the barrage in their deep concrete bunkers and, warned by the cessation of the bombardment ten minutes before zero hour (another costly error) they emerged to the sound of bugles from below ground and reached their machine guns before the slowly advancing British troops were even halfway across no-man's-land. In those areas where the British did succeed in taking German positions, they found them in-

[60] M. Middlebrook, *The First Day of the Somme,* p. 161.

tact and even with the electric lights still working. So much for Rawlinson's crushing barrage.

The British troops encountered not just rifle and machine gun fire, they also marched into the face of a counterbarrage from the German artillery, which had similarly survived Rawlinson's bombardment. Even where the wire had been broken the passages created served as death traps, for the Germans had concentrated their firepower on these openings. Elsewhere, the wire trapped thousands of men, equipped only with pitiful hand-cutters, who could find no way through and milled about like flocks of sheep until the machine guns scythed them down. One battalion of the Newfoundland Regiment suffered ninety-one percent casualties on the wire.

The regiments which suffered worst contained the "Pals" battalions. As wounded men began to stream back, anxious battalion commanders in the second wave telephoned for new orders but the answer was always, "You must stick to your plan. You must carry out orders." On the left of the advance—around Gommecourt and Beaumont Hamel—the British made no progress at all and suffered dreadful casualties. In the center, behind which Gough's Reserve army of three cavalry and two infantry divisions was massed, some progress was made, but successes were isolated. Even where German trenches were taken the forward troops could not communicate with the British lines to tell them of their success as runners were shot down in hundreds and telephone wires cut by artillery fire. Heroism was present all along the line, notably by the 36th Ulster Division, which took the Schwaben Redoubt in front of Thiepval and held it without reinforcements until driven out by German counterattacks, having suffered over 5,000 casualties. Only on the right, where better French tactics prevailed, was real progress made. The Germans were taken by surprise here, for they had not expected the French to be able to launch an attack in view of their recent martyrdom in the great struggle around Verdun.

By the end of the day the British held a three-mile-wide portion of the German position, to a depth of one mile, and just three of the thirteen target villages. At no point had they reached the second line of German defenses. Haig's breakthrough was out of the question—the cavalry would not be needed. Rawlinson had more reason to be pleased than Haig—his men were at least "holding" a small sector—but it had been achieved at the cost of nearly eight casualties for every one German. Haig blithely commented that the casualties could not be considered severe in view of the numbers engaged. Yet fifty percent casualties were so rare in military history that they usually represented a defeat as decisive as the French had suffered at Waterloo. No previous army commander with such casualties had ever expressed himself satisfied. Haig's complete ignorance of events was clearly demonstrated by his unforgivable attribution of cowardice to Hunter-Weston's Eighth Corps. In Haig's words, "few of the VIII Corps even left their trenches." In fact, Eighth Corps suffered over 13,000 casualities—the highest of any corps involved. And what was awaiting the casualties who were lucky enough to be brought back from the front? Between Albert and Amiens a casualty clearing station had been set up to expect 1,000 casualties. Within a few hours of the start of the battle they were overwhelmed by 10,000 wounded men. A surgeon there wrote of, "streams of ambulances a mile long" waiting to be unloaded.

The five months of fighting on the Somme saw the most bitter attritional battle not just of the First World War but in all history. The British and the Germans fired over thirty million shells at each other and suffered in return over a million casualities in an area little more than seven square miles in extent. It was the biggest abbatoir ever devised by man.

9

The Battle of Caporetto, 1917—A Monologue in One Act

It is doubtful if Erwin Rommel ever read much poetry by Rudyard Kipling. Had he done so he might have paused for a moment over the lines from the poem "If":

If you can meet with Triumph and Disaster
And treat those two impostors just the same . . .

For the life of a soldier is intimately tied up with the impostors "triumph" and "disaster." And Rommel rode a switchback during his military life, experiencing triumphs in France and North Africa and yet living long enough to meet disasters not far from the scenes of his greatest achievements.

As a young officer, replete with the arrogance of youth, Rommel played a part in one of the great "triumphs and disasters" of the First World War—the battle of Caporetto—for it was both: triumph for the Central Powers and disaster for Italy. While historians mapped the passage of armies, even whole peoples, individuals like Rommel stood out like fixed stars in the firmament. The part he played in the drama has no doubt been exaggerated. Yet, exaggerated or not, Rommel's achievements were a product of a spirit born of total self-confidence and absolute conviction. Those who "held the spears" were merely his props in the theater of war. The Italians—his victims—were almost willing sacrifices at his altar. By 1917 their fighting spirit had gone; their belief could only be rekindled by a man of his egotism if not his ability. It was heady stuff for any young officer, let alone an embryonic military genius.

In 1917, the German High Command turned its attention to the Italian Front, deciding to help their Austrian allies knock Italy out of the war with one tremendous blow. Field Marshal Eric von Ludendorff was sent to oversee a great offensive in the Dolomites, which would be spearheaded by fifteen infantry divisions, including the eight elite German divisions of General von Below's Fourteenth Army. The Italians were in for a shock. The cosy—if bloody—war against their fellow incompetents the Austrians was about to be taken by the scruff of its neck and shaken. Operation *Waffentreue* (Brothers-in-Arms) was about to burst on them.

In the early hours of 24 October, 1917, the battle of Caporetto began with a volcanic bombardment from General von Below's Fourteenth Army, mixing high explosives with gas shells. The Austrians knew well that the Italian gas masks were quite inadequate and realized just how effective gas would prove on this front. They could advance, confident in the knowledge that large numbers of Italian troops would be incapacitated. Typically there was no Italian counter fire—one officer even refused to use his guns in order to conserve ammunition—and the Austrians knew from long experience that there would be none. They also knew that once the Italians suffered a setback they would collapse like a house of cards. They were right. The Italian commander-in-chief, General Cadorna—controlling the battle from Udine, twenty miles from the fighting—decided he would be safer farther back and retreated a hundred miles to Padua. It was disgraceful leadership and it set the pattern for generals, officers, soldiers, and civilians to run away. The problem was that Italian morale was at rock bottom. Numerous murderous Italian assaults against the Austrians along the river Piave had produced no results over a period of almost two years. Now that the German High Command had decided to send both troops and a general to help their Austrian allies, the Italians rightly feared that they were in danger of total defeat.

As the Austrians drove the Italian Second Army back, morale collapsed among the Italian troops, who shouted "blackleg" at reserves moving to the front. Austrian advanced units met Italians shouting *"Eviva La Austria."* Disintegration followed, with most of the Second Army heading home to get into civilian clothes as quickly as possible. Whole Italian divisions disappeared from the front, along with their generals and staff officers, while their artillery was only notable for being in the forefront of any scramble to escape. But as the Italian army fled south, vast numbers of refugees—as many as 500,000—fled with them, blocking all the roads with their horses, carts, and other paraphernalia. In fact it was these refugees who turned out to be Italy's best soldiers, holding up the Austro-German advance more effectively than Italy's army had done. As they fled, the refugees descended on peasant villages like a swarm of locusts, looting and taking all the food and livestock. The Austrian advanced troops found not just the roads but the fields alongside packed with millions of people, soldiers and civilians, so that it was impossible to take any more prisoners.

The Italian reverse at Caporetto was an astonishing event; more than a military defeat, it took on all the appearances of a major natural disaster, affecting a whole population. It was the most stunning victory of the entire First World War. The elimination of the Italian Second Army, combined with terrible losses elsewhere, meant that the Italians suffered 800,000 casualties, mostly from desertion. For a loss of just 5,000 men the Austrians and Germans had scored what should have been a war-winning victory. What saved Italy was the inability of the enemy to exploit it. Ludendorff and von Below had instigated *blitzkrieg,* but they lacked the Stukas and the fast-moving tanks of the Second World War to make their breakthrough decisive. Eventually, with a stiffening of British and French divisions from the Western Front, the Italians established new positions on the river Piave, and the panic subsided. But for a while it seemed as if an avalanche had streamed from the mountains to crush Italy.

Rommel marched to Italy as a captain in the Württemburg Mountain Battalion, which had seen heavy fighting in the short war against Rumania in 1916. Stoicism seemed to be the order of the day in Rommel's unit, and fighting spirit was strong as a result of a sense of duty rather than any elation on the part of the soldiers. Rommel simply took everything the Italians could throw at him and shrugged it off. He expected nothing less from war, and the Italians were far less formidable opponents than he had expected. As a historian wrote of the Austro-German forces at this moment:

> The scent of victory is an extraordinary stimulus. It dissipates the misery of wet, cold and mud, and the fatigue of carrying heavy packs. It can turn mediocre troops into good ones, though only for a short time; they have to be well-trained and well-led if they are to survive these handicaps for long. Only then will the flame set alight in them continue to burn.[61]

Rommel's troops expected nothing less than victory over the Italians and this gave them an enormous moral advantage in the fighting ahead. Underestimating the enemy can be one of the most inadvisable approaches any commander can make. However, Rommel would have argued the case for realism. The Italians were inferior; to credit them with more ability would be to counsel caution where dynamic action was the far better course.

Once in the mountains, Rommel moved with lightning speed and little respect for his senior commanders, who advised caution. Near the mountainous Matajur Ridge, Rommel spotted large numbers of Italian troops crowding together indecisively in the face of the German advance. Their spirit seemed broken and so Rommel walked toward them alone, waving a handkerchief. With astonishing willpower he called on them—there were at least 1,500 of them, with forty-three officers—to surrender. Had he shown the slightest indecision he would have paid with his life. Instead, when he was no more than 150 yards from them, the Italians flung aside their weapons and ran toward him, lifting him on their shoulders and shouting, *"Eviva*

[61]C. Falls, *Caporetto 1917*, p. 64.

Germania!" It was an astonishing achievement by a single officer in the face of such odds. One Italian officer who refused to surrender to Rommel was immediately shot down by his own men.

Any other man might have thanked the fates and left someone else to mop up the remainder of the Italian Army, but not Rommel. Ordered to fall back by battalion headquarters, Rommel decided to disobey orders. Even more convinced of the superiority of his German troops over the Italians, Rommel ordered his mountain troops to encircle the 2nd Regiment of the Salerno Brigade, which was holding the hills nearby. Yet again Rommel decided he needed no support. Ahead of him, on the winding mountain road, he had spotted another large group of Italian soldiers, from the Salerno Brigade. They seemed ready to lay down their arms and so he tried the same bluff that had worked earlier. He noticed the regimental commander seated on the road, with several of his officers beside him. Rommel promptly walked up to the group of Italian officers and ordered them to move away from the rest of their men. Then he ordered the Italian rank and file, amounting to nearly 1,200 men, to march swiftly down the road into the waiting arms of his men. While all this was happening the Italian officers simply watched in tears. It was as abysmal a display by them as it was an incredible one by Rommel. By the end of the campaign, Rommel's unit—and that usually means the man himself, as he was enough of a *prima donna* to share his glory with no man—claimed to have captured 150 officers and nearly 9,000 Italian troops. Rommel's apparently magical success was due to his willpower. He had gained a moral advantage over the Italian officers, who expected the worst and tended to receive it.

For those readers unconvinced by this demonstration of the role of the individual in history, it is as well to point out the fact that the Italians had no quarrel with the Germans and many of them had no wish to pursue an unpopular war. This presumably may have contributed to the ease with which Rommel accepted the surrender of such large numbers of enemy troops. Politicians in Rome or Milan might have sought to benefit by an alliance with Britain and France, but the Italians of the Dolomites were not committed to the fight. Italy's war aims were entirely territorial. When a man's life is at stake, whether or not Italy rules in Fiume or the Dalmatian Coast may seem to count less than whether he ever returns to see his wife and children. Few Italians were career soldiers like Rommel. The great man received the *Pour la Mérite* for his extraordinary actions. Yet as one reads of his antics in this campaign, one feels that just occasionally some mention might have been made of the men who helped him, the common Württemburgers without whom nothing would have been possible. But then, of course, modesty was not one of Erwin Rommel's military virtues.

10

"With Our Backs to the Wall"—Ludendorff's Offensive, 1918

Throughout history generals have been, perhaps, rather too fond of asking their troops to "fight to the last man." But what exactly are they asking of their men? Do they expect a kind of Custer's Last Stand, with the soldiers forming a ring around the last officer standing? This may have been appropriate in the colonial wars of the nineteenth century, where the enemy was generally disinclined to observe the finer points of the Geneva Convention. White troops usually had little alternative but to fight to the bitter end, saving the last cartridge for themselves. Zulus, Ashanti, Dervishers, Sioux, and Cheyenne were not noted for taking prisoners. However, in the wars between so-called civilized powers, it has usually been possible to save one's life by surrendering at an appropriate moment. However, some generals—often at the instigation of political masters—have felt that they could increase their soldiers' fighting spirit by ordering them to die at their posts. This demand may sound impressive as part of a political speech—one has only to remember Winston Churchill's rallying call to the British garrison at Singapore, as the Japanese closed in on the city—but it is rarely taken seriously by the soldiers who have to accept death as part of their duty. Like Churchill in 1942, Hitler called on his troops in Stalingrad to die at their posts. He instructed General von Paulus to fight to the last man, even promoting the general to field marshal on the grounds that no previous German field marshal had ever surrendered. Hitler soon found that there was a first for everything as von Paulus took the only step possible and came to terms with the Russians. Soldiers will obey their orders until they perceive that their orders no longer have any relevance. At the last moment every human being will be forced back on his prime directive of self-preservation, and if this involves surrendering to the enemy in order to preserve his life, most soldiers will take that option. Where indoctrination has been very strong, as in the case of the Japanese, with their adherence to the theory of *bushido,* soldiers may prefer suicide to capture, with all its consequent dishonor. The ancient Romans often preferred death to dishonor, because in many cases—as with the German tribes of the Teutoburger Wald—their opponents were so cruel and barbaric that death by one's own hand was preferable to a prolonged death by torture as administered by the enemy. Otherwise, surrender does have its attractions, as the British Fifth Army found on 21 March, 1918.

Victory on the Eastern Front in 1917 gave the German High Command the opportunity to transport nearly a million extra soldiers to the Western Front where, for the first time in the war, they would hold a numerical advantage over the Anglo-French forces. By 1918 the Germans were convinced that even a victory over the French army, which had been unsettled by mutinies during 1917, would not persuade Britain to sue for peace. Only a decisive victory over the British army would give Germany the chance of victory before American troops arrived in such numbers as to swing the pendulum back to the Allied side. As a result, Field Marshal von Ludendorff planned a gigantic offensive against the already-depleted British Fifth Army, commanded by General Sir Hubert Gough, between the rivers Oise and Sambre, at the point where the British and French armies joined. Ludendorff's planning was excellent. There were strong reasons to suggest that the Fifth Army would succumb to a sudden *blitzkrieg* attack, prefaced by a short but overwhelming artillery bombardment and carried out by fast-moving elite "shock" troops, using new tactics and advancing behind a "creeping" barrage. The Fifth Army was, in fact, the weakest of the British armies and covered the longest stretch of front. Ludendorff believed that British leaders had grown complacent, relying on the perceived historical "invincibility" of British troops. Once the Fifth Army had been shattered and British troops forced to retreat in disorder, the Germans believed Britain would be far more willing to entertain peace negotiations. Moreover, the collapse of their British allies in the north would persuade the French to fall back to defend Paris, allowing the Germans to reach the Channel coast and drive the British into a full-scale evacuation of the kind that was to take place at Dunkirk twenty-two years later.

The Germans had not been able to understand how the British army had survived the series of offensives that it had undertaken in the latter half of 1917 without cracking. Field Marshal Haig's command had suffered over 800,000 casualties in the previous twelve months and was nearly 70,000 men below establishment. Even the German army had experienced internal tensions and a loss of morale. In fact, only the German adoption of the defensive in 1917 had prevented serious disruption. Meanwhile, the French, Italian, and Russian armies had all experienced disorder and even collapse through exhaustion in the same year. However, as the Germans suspected, the British army was on the brink of collapse even as it achieved its most hollow of victories at Passchendaele in November 1917. Combat exhaustion was at an unparalleled level during the winter of 1917–18 and, as more than one French general had observed in previous centuries, if the British had known how badly damaged they were, they would have been defeated. Stubbornness and an unshakable will to win prompted Britain's generals and soldiers to go on fighting at whatever cost. Yet the fighting in the muddy slime of Passchendaele had filled the hospitals with thousands of men who were mentally rather than physically destroyed. British doctors, with little experience and even less sympathy for shell shock, frequently classified such patients as "Not Yet Diagnosed," as if once the solution had been found, the men could be packed off back to the trenches to endure even more of the same. One patient who recovered sufficiently to return to the trenches was described as:

> a man with a very inferior physique. He lay with his head under the blankets. When his head was uncovered, his gaze became fixed and he had the appearance of witnessing some terrifying spectacle. This lasted for a few minutes, and then he broke into loud weeping. He appreciated questions only when shouted at loudly. He could not speak, sit up or walk. When placed on his feet he collapsed limply.[62]

There must have been thousands of men just like this in all of the combatant armies on the Western Front at this time. The British Fifth Army alone had 8,000 men suffering from this "exhaustion psychosis" during the winter of 1917–18. Others suffered from amnesia or unbearable memories of their time in the trenches, like this one described in the British Official History.

> He kept up a constant low muttering, snatches of which were intelligible and were found later to refer to his experiences in the line . . . He appeared to be living again and again through his recent experiences . . . and interpreted anything occurring near him from this point of view. Those next to him he thought were comrades in his unit whom he addressed by name and talked to continually; or they were German snipers and so on. Frequently he would sit up in bed with a terrified look on his face, shout "Look, look!" after which he would sink back in bed moaning a friend's name.
>
> Any noise occurring near was thought to be from bombs, shells, machine guns or other weapons, and was liable to bring on one of these brief outbursts.[63]

Only some form of *esprit de corps* or, at the last gasp, simply self-respect and unwillingness to be seen to fail in front of others kept many men in the line. Sergeant French spoke for many when he described his own feelings at this time:

---

[62]W. Moore, *See How They Ran,* p. 29.
[63]Ibid., p. 29.

> My fright was such that I was conscious of biting on the stem of my pipe to prevent my teeth from chattering. If I had been alone on that occasion I believe that I should have been cowering at the bottom of the trench in hopeless terror; it was the presence of others that saved me. Just as in boyhood I had acted a part, trying to appear natural when walking across a room, so now I was acting again, counterfeiting bravery.[64]

Sadly, British attitudes toward shell shock were far from progressive. A severe manpower shortage was making it necessary to return wounded or mentally disturbed patients to the front far too soon. Britain was simply overcommitting herself, with massed armies fighting the Turks in Mesopotamia and Palestine, the Germans in various parts of Africa, the Bulgarians in the Balkans and bolstering the Italians after their collapse at the battle of Caporetto. In addition, the Royal Navy, swollen by wartime conditions, absorbed another million men. British medical officers, presumably sometimes against their own better natures, turned a stern face against the whole issue of combat fatigue. The result was that morale in the British armies on the Western Front declined as more and more young boys and crippled old soldiers were moved up to the front line to face the Germans. William Moore records the callous treatment accorded to one invalid—an officer at that. This tunneling officer had been trapped underground after the Germans exploded his mine prematurely, and he suffered gassing as well. As a result the man suffered a complete collapse and total amnesia. He was evacuated to England with the diagnosis, "his aspect and regard have those of extreme mental defect." He was unable to perform even the simplest toilet act for himself and was unable to feed or dress himself. With care the man might have made a partial recovery but after six months in England he was passed fit by a medical board and returned to France, for more service in a tunneling unit.

By the end of the Passchendaele campaign, in November

[64]W. Moore, *See How They Ran*, p. 31.

1917, there was scarcely a man who did not jump at his own shadow. Norman Gladden described how easily panic could break out when through exhaustion men had lost the power to control their fears:

> As a line of Khaki figures appeared along the horizon from the rear, pandemonium seemed to be let loose. Coloured lights dropped from the plane, while in front fearful red rockets shot into the sky. The relief was in but the enemy no doubt thought that another attack was developing. "Clear out" and "Every man for himself" were shouted along the line. Panic took command. In a mad stampede we passed through the relieving company. Undoubtedly we had lost our nerve . . . I saw men crying and I would have cried myself had I the tears. The company that night was in the grip of a sort of communal terror, a hundred men running like rabbits.[65]

Just occasionally true relief was found, often by chance. When the British 23rd Division was transferred from Ypres to the Italian front, it was taken by rail through the unspoiled countryside of France, leaving the horrors of war far behind. As one officer wrote, "Our morale has increased 50 per cent today; the Riviera has revived our battered spirits as much as a tot of rum would have done. I feel quite tired with rushing from one side of the carriage to the other to see some new and wonderful sight." But such an escape was available to few British soldiers at the start of 1918. Exhausted by their efforts in 1917, they were now to face an overwhelming attack by dozens of fresh German divisions. And as the cry for reinforcements went up, in Britain the politicians became increasingly loath to allow Field Marshal Sir Douglas Haig, the British commander-in-chief, to squander British lives on more attritional battles. Prime Minister David Lloyd George held back more and more men of military age. Significantly, War Office figures for 1 January, 1918, reveal that there were 38,225 officers and 607,403 fit and fully trained sol-

[65]W. Moore, *See How They Ran,* p. 32.

diers in Britain, who could have gone to France and stiffened the British armies, so that a German breakthrough might have proved impossible. Haig's armies were being starved just at the moment of their greatest danger. In addition, men who previously would have been conscripted and rushed to France were being given deferment or being transferred to "vital" war industries. As some cynics observed, the British, having bled France dry of her men, were now awaiting the arrival of America's soldiers to win the war for them. Meanwhile, the British navy—at minimal cost—maintained her hold on Germany's windpipe.

At the start of 1918, the British became aware that their real enemy was time. The American troops were arriving in France at a very slow rate, while the Germans were transporting ten whole divisions per month from Russia. Gradually, the German High Command—now the duo of Hindenburg and Ludendorff—were amassing such local superiority in Flanders that they would be able to launch an offensive against any point in the British line with a real chance of success. In fact, Ludendorff was planning a knockout blow of Napoleonic finality. Three whole armies would attack the British Fifth Army and the initial breakthrough would be spearheaded by new mobile squads of storm troopers. To maximize the efficiency of the artillery barrage that would precede the attack, Colonel Bruchmüller, master of the creeping barrage, had been moved to the Western Front. It was an ominous sign for the British and should have warned them of what to expect. In a sense, the British knew already what was coming, but simply could do nothing about it. In terms of quality of troop and fighting spirit they were—at least temporarily—thoroughly outclassed by the Germans.

The storm burst at 4:00 A.M. on 21 March and the great offensive that was launched is sometimes known as the Kaiser battle, because Kaiser Wilhelm II moved up to army headquarters to personally involve himself in what he supposed would be the war-winning victory. Along fifty miles of the British front line between the rivers Oise and Sambre 6,000 German guns opened a five-hour barrage that was the biggest the world had ever known. Of the 100,000

British troops facing this German onslaught, surprisingly few succumbed to the shells—no more than 2,500 were killed and 6,000 wounded—but it was not merely death and destruction that were the Germans' aim, but demoralization. In this they were very successful. When the barrage was halted, the German infantry poured forward through the huge hole that the artillery had blown in the British defenses.

In contrast with the depressed and exhausted British, the German troops marched forward with supreme confidence, convinced that the end of the war was in sight. German morale was very high. As Private Wilhelm Boscheinen shows, the call of duty was very strong:

> Feelings before the battle? Can I describe them after so many years? The older men couldn't care less, only to be out of the shit. The younger ones of course are frightened and anyone who dares to deny it is a liar. But also the tension. How will it go? But at that time we were brought up through school and parental discipline in the spirit of the military empire of the Kaiser.[66]

Private Adolf Vogelsang shared the same sense of duty, yet underlying it was a feeling that it was a "poor man's fight but a rich man's war":

> A time like this was bound to create a certain anxiety. Having been under fire again and again in the past, it wasn't too bad for me. The call of duty and iron discipline worked all right. One thinks about home and the family and God. Didn't all people expect help from Him? Hadn't they been told they were fighting for a just cause? Didn't they all pray for victory and peace? How often His name has been misused to start a war, not in the interests of the people, but in the interests of ambitious and greedy rulers of the land and the church?[67]

[66]M. Middlebrook, *The Kaiser's Battle,* p. 168.
[67]Ibid., p. 168.

Nevertheless, patriotism was still strong among the Germans, even after four years of war, as Lieutenant Hermann Wedekind shows:

> Just before the bombardment ended, the battalion commander, Major Scherer, started to sing *"Deutschland, Deutschland über Alles"* and the singing spread across to us and we all joined in. It was the first time that I heard our men singing the National Anthem since the autumn of 1914 when our young volunteers had been heard singing it in their attacks in Flanders. The spirit now wasn't the same as in 1914 but I think the battalion commander sang this to take our soldiers' minds off the coming battle. The men were quite happy—there was no English artillery fire and we thought that there were no English soldiers left alive.[68]

Inside the first sixty minutes General Gough had lost more than thirty percent of Fifth Army's infantry strength. Almost all of Gough's divisions had been through the Passchendaele bloodbath and they could not stand the new German bombardment. For many men of the Fifth Army the German barrage marked the final collapse of their morale. As Private Atkinson wrote:

> Artillery was the great leveller. Nobody could stand more than three hours of sustained shelling before they start falling sleepy and numb. You're hammered after three hours and you're there for the picking when he [the Germans] comes over. It's a bit like being under an anaesthetic; you can't put a lot of resistance up. The first to be affected were the young ones who had just come out. They would go to one of the older ones—older in service that is—and maybe even cuddle up to him and start crying.[69]

Some British units kept fighting while others ran. The German storm troops seemed so elated by the success that they showed uncharacteristic sympathy for the British

---

[68]M. Middlebrook, *The Kaiser's Battle,* p. 169.
[69]Ibid., p. 161.

troops as they overran them. As Private Parkinson found, some encounters with the enemy could be almost surreal:

> We were in action for some time and I think we hit many Germans. Then it went quiet and I thought that we had stopped them. I was loading another belt into the gun when I felt a bump in the back. I turned round and there was a German officer with a revolver in my back. "Come along, Tommy. You've done enough." I turned round then and said, "Thank you very much, sir." I know what I would have done if I had been held up by a machine-gunner and had that revolver in my hand. I'd have finished him off. He must have been a real gentleman.[70]

Some British officers were not keen to retreat. But Private Thomas Link was not eager to throw his life away because his colonel was a hero:

> Another runner had come down from the Yorks and Lancs to our relay post between the forward positions and battalion headquarters and he told our colonel that the message was that we had to retreat. Colonel Smith said, "There is no such damned word in the British Army as 'retreat.' I've lost one eye for one medal and I'll lose the other for the V.C." Then he told me to come with him. He led the way with a revolver in each hand. I followed him down this trench and then we reached the junction with a trench from the left which Gerry had covered. As we went past the opening of this, a bullet came *ping* in between us. The Colonel said, "That's bad aim on your part, Gerry." I let him go on a few more yards down the trench towards where we could plainly hear the fighting. I didn't say a word to him but turned back. I knew I might be shot as a deserter but I would have been shot or taken prisoner if I had gone on. I went back to where I had started from but the others there had gone. I was left on my own but I caught them up later.[71]

---

[70]M. Middlebrook, *The Kaiser's Battle,* p. 192.
[71]Ibid., p. 195.

As might be expected British resistance varied considerably from unit to unit. In his book *Storm of Steel* Ernst Junger describes the different responses he encountered. In some places "the British emerged with arms uplifted and knocking knees," while in others "the fellows put up a superb show." Corporal Willy Adams noted much the same: "There were a few parties of English in scattered positions who didn't fight very hard. They threw away their rifles and wanted to surrender." Yet, however well or badly the British fought the result was the same: a complete collapse of the British front line in front of Péronne. British observers were in complete agreement as to what had happened. Captain Essame of the Northants Regiment described something he had never expected to see: a British rout:

> I got the impression that the Army had definitely been broken. We didn't even bother to try to rally the stream of fugitives coming across the bridges. An awful lot of them had had enough. They were making for somewhere way back.
>
> For the first time I saw British soldiers coming back without their rifles. They weren't men of labour battalions or non-combattant troops. They were men from infantry regiments. I'd seen men retiring on the Somme and at Passchendaele, but they had always brought their rifles with them. This was the first time I'd seen men coming back without them.[72]

Lieutenant Richard Gale was marching at the head of reinforcements when he encountered a similar fugitive mob:

> We heard no fighting, nor had we seen any formed body of troops; but what we had seen appalled us. Dumps of kit and valises lay on the side of the road, disorganized transport and guns were moving to the rear, all intermingled with pathetic groups of refugees. Canteens had been abandoned and their stores of spirit rifled. This was a retreat with all the horrors of panic. There was, as far as we knew,

72W. Moore, *See How They Ran*, p. 84.

> nothing between us and the Channel ports, save this wretched rabble that seemed to have lost all cohesion and will to fight.[73]

The British artillery had tried to match the German guns, but they were simply overwhelmed. Nevertheless, they stuck to their task and fired off their ammunition. Far behind the front line they were unaware of the collapse until soldiers began pouring back in confusion. Lieutenant Knee witnessed an incongruous sight on this day of disaster: "Then we met some more gunners, belonging to a battery of 6-inch howitzers, playing football—out of ammo, I suppose."

Elsewhere, some regular officers were trying to stem the flight to the rear. Lieutenant Colonel Sir Ian Colquhourn of the Scots Guards was marching up and down with pistol in hand like a housemaster trying to stem a horde of unruly schoolboys. Colquhourn was a much-loved figure in the army and was known as a "soldiers' man"; he had once been court-martialed by General Haig for arranging a "Christmas Truce" with the Germans against orders. Not far away Regimental Sergeant Major Withers—known to his men as "African Joe" from all his Boer War medals—was calling on the retreating soldiers to "Be British! Be British!" and turn back to face the enemy.

But panic can take many forms. Some men panicked that day by running away, others by fighting when there was no hope. Many individual soldiers, even some units, took it upon themselves to fight to the death. German progress was not uniformly easy. Near Beaumetz, the Germans encountered ferocious resistance from elements from the Seaforth Highlanders and the Gloucestershire Regiment. William Moore recounts an incident displaying the courage of desperation. The Germans had been held up by what they called the "pigheaded Scotsmen" and in front of the British position the German dead lay piled. But at last they managed to turn the British flank and bring up machine

[73]W. Moore, *See How They Ran,* p. 87.

guns, some even in the rear of the Seaforths. There was no hope now. With ammunition running out, it was only a matter of time before the position was overrun. At last a giant Seaforth captain cracked. Seizing a rifle he leaped onto the parapet and alone and in full sight of the Germans he began firing wildly at them. At that moment he was flattened by a flying rugby tackle from one of his colleagues, a captain of the Gloucestershires. "You're a bloody hero, but also a bloody fool," said the Englishman to the Scotsman, on whom he was sitting. There was no need at that moment for the brave Highlander to throw away his life. That was not courage but desperation. Within minutes a German officer made his way down the trench toward the two British officers. He was carrying a stickbomb, but instead of throwing it he saluted and asked them to surrender. Honor had been served. Nothing was to be gained by useless resistance.

Unfortunately, Lieutenant Colonel J.H. Dimmer, V.C., M.C. would not have agreed. As Martin Middlebrook says, Dimmer was a "Blood and thunder" soldier, who came from a "blood and thunder" family. He was the worst/best [depending on your viewpoint] kind of officer for this sort of crisis. At the outbreak of war in 1914 he had been serving in West Africa, but he immediately sailed for home, joined his regiment and fought at the first battle of Ypres in 1914, where he won the Victoria Cross for outstanding courage in staying with his machine gun and continuing fighting even though he had been wounded five times. On 21 March, 1918, he was in command of his own battalion of Berkshire Territorials and his men were proud to be part of "Colonel Dimmer's Battalion." But there are times when common sense rather than courage is required, and this was one of them. Dimmer only knew how to lead from the front. He therefore determined to lead his men in a counterattack against the advancing Germans. But instead of concealing his officer's regalia (that made him an easy target) and leading his men from a crouching position, Dimmer insisted on mounting his horse so that everyone could take courage from his example. His officers tried to persuade him to dismount, but he was not a man for turning. With his groom riding by his side, he set off toward the Germans, with two companies of his men stumbling behind him. Soldiers from other regiments watched the amazing scene. A Gloucestershire private remarked: "We were astonished; we just couldn't believe it. It was good riding country but not in those conditions." Another observer commented, "We realised that the two horsemen were silhouetted against the skyline and we put up covering fire to protect them. But as soon as they reached the top, they were picked off and fell to the ground." Dimmer was no El Cid, unfortunately. His death was futile. Even as he had set off on his last ride, the Germans were planning an attack of their own. In the event, forty of his men were caught unnecessarily in the open and killed, and the rest of his battalion cracked and took to their heels, shocked by the sudden reversal of fortune that had cost them their beloved leader and the security of their trenches.

In simple terms the British Fifth Army had cracked and had run. Their attitude was well summed up by a nameless private who wrote to Martin Middlebrook fifty years after the event: "When the Jerries came towards our line in large numbers, they were firing from the hip and I thought, 'Tosh. Do what some of the others are doing. Hop it back.' So I did. I was not alone, I can assure you, otherwise I don't think I should be able to write this." There was no disguising the fact that British resistance had been patchy. Captain Slack of the East Yorks was honest about his own men:

> There was a tremendous amount of confidence in most of the men, but there were young boys, young men who'd just come out and who hadn't got the *esprit de corps,* and they were running away. They didn't care a damn. They didn't mind anything. Rifles gone. I actually had to draw my revolver on an officer.[74]

And when the call came for all good men and true to come to the aid of their country, there were many who

[74]W. Moore, *See How They Ran,* p. 120.

lacked the spirit. One regular sergeant major, an old soldier, could not understand the lack of fighting spirit among his fellow countrymen in the days that followed the great retreat:

> You never saw such bastards. There'd be a notice put up in the morning that all men of the 3rd and 18th Divisions would parade at twelve, and for the 36th and 56th Divisions at five. When the time came there wasn't a man of these divisions to be found. Oh no, they all belonged to the 57th and 47th. And next day when the parade was for the 57th and 47th, they all belonged to the 3rd and 18th Divisions. They tore the badges off their jackets and lay doggo.[75]

All this seems a far cry from the enthusiasm with which millions of Britons volunteered to fight for their country between 1914 and 1916. But disillusionment with the war was profound by April 1918. Without conscription, it is doubtful if any of the warring nations could have kept an army in the field. Moreover, for the first time, it was beginning to dawn on some of the stubborn English that they might actually be going to lose the war. On the other hand, the Germans were beginning to fall victim to their own success. The Germans were undoubtedly motivated—as no previous army had been in the war—by the expectation of loot. Sergeant Friedrich Flohr explained: "We knew that the Tommies had in their dugouts all the good things that we hadn't—chocolate, coffee, corned beef, wine, spirits, cigars, cigarettes." Sergeant Hermann Gasser agreed with him, "Our personal view about the attack wasn't so much concerned with the English; what we had much more in mind was the booty—the provisions, stores, cigarettes, tinned meats, biscuits. We knew what we were after."

As the Germans moved through the British positions it was as if an army of beggars had been let loose in a supermarket. They were simply amazed at the luxuries they found there. After the privations they had endured it was like Christmas. Corporal Willy Adams wrote:

[75]W. Moore, *See How They Ran*, p. 125.

What we found was a gold mine for hungry soldiers—things we hadn't seen for years. What a difference from our food! We just stuffed ourselves. I found a tin with a 100 cigarettes; they were the best I have ever smoked in my life. We opened every tin in sight because none of us could read English. I especially remember a tin with baked beans and pork. I enjoyed that very much.

I next explored a dugout that had apparently recently been abandoned by English artillery officers. There was an enormous gramophone on the table that Haller at once set going. The gay musical-comedy song that whirred from the disc made a ghostly impression, and I threw the box to the ground, where after a wheeze and a gasp it lay still. The dugout was furnished with extreme comfort, even to a little open grate and a mantelpiece, on which lay pipes and tobacco, with a circle of armchairs round the fire. Merry old England!

Naturally, we took without compunction whatever we liked. I chose a haversack, underclothes, a little silver flask full of whisky, a map-case, and some most charming toilet articles by Roger et Gallet, no doubt tender recollection of a Paris leave.

A neighbouring room served as the kitchen, whose array of provisions filled us with respectful admiration. There was a whole boxful of fresh eggs. We sucked a large number on the spot, as we had long since forgotten their very name. Against the walls were stacks of tinned meat, cases of priceless thick jam, bottles of coffee-essence as well, and quantities of tomatoes and onions; in short, all that a gourmet could desire.

This sight I often remembered later when we spent weeks together in the trenches on a rigid allowance of bread, washy soup, and thin jam. For four long years, in torn coats and worse fed than a Chinese coolie, the German soldier was hurried from one battlefield to the next to show his iron fist yet again to a foe many times his superior in numbers, well equipped and well fed. There could be no surer sign of the might of the idea that drove us on. It is much to face death and to die in the moment of enthusiasm. To hunger and starve for one's cause is more. . . .[76]

[76]M. Middlebrook, *The Kaiser's Battle,* p. 213.

The problem for the officers was that the plundering began to affect discipline, as Rudolf Binding found:

> Today the advance of our infantry stopped near Albert. Nobody could understand why. Our airmen had reported no enemy between Albert and Amiens. The enemy's guns were only firing now and again on the very edge of affairs. Our way seemed entirely clear. I jumped into a car with orders to find out what was causing the stoppage in front. Our division was right in front of the advance, and could not possibly be tired out. It was quite fresh. When I asked the Brigade Commander on the far side of Meaux why there was no movement forward he shrugged his shoulders and said he did not know either; for some reason the divisions which had been pushed on through Albert on our right flank were not advancing, and he supposed that this was what had caused the check. I turned round at once and took a sharp turn with the car into Albert.
>
> As soon as I got near the town I began to see curious sights. Strange figures, which looked very little like soldiers, and certainly showed no signs of advancing, were making their way back out of the town. There were men driving cows before them on a line; others who carried a hen under one arm and a box of notepaper under the other. Men carrying a bottle of wine under their arm and another one open in their hand. Men who had torn a silk drawing-room curtain from off its rod and were dragging it to the rear as a useful bit of loot. More men with writing-paper and coloured notebooks. Evidently they had found it desirable to sack a stationer's shop. Men dressed up in comic disguise. Men with top-hats on their heads. Men staggering. Men who could hardly walk.
>
> They were mostly troops from one of the Marine divisions. When I got into the town the streets were running with wine. Out of a cellar came a lieutenant of the Second Marine Division, helpless and in despair. I asked him, "What is going to happen?" It was essential for them to get forward immediately. He replied, solemnly and emphatically, "I cannot get my men out of this cellar without bloodshed." When I insisted, assuming from my white dragoon facings that I belonged to the same division as himself, he

> invited me to try my hand; but it was no business of mine, and I saw, too, that I could have done no more than he.[77]

Meanwhile, the authorities in England, who had held back reinforcements from their generals, now awoke to the fact that their action might be going to cost Britain the war. Immediately, replacements were rushed to France and in the space of a week over a hundred thousand new men arrived at the front. However, "men" might be an exaggeration. One Australian observer—the Australians were never impressed by the physiques of Britons from the industrial cities—wrote dismissively of the new arrivals:

> For two days companies of infantry have been passing us on the roads—companies of children, English children; pink faced, round cheeked children, flushed under the weight of their unaccustomed packs, with their steel helmets on the back of their heads and the strap hanging loosely on their rounded baby chins.[78]

Exaggerated as this obviously is, it was difficult to expect such young soldiers to step into a crisis of such proportions and stem the tide of German advance. For a number of reasons, the fighting spirit of such young men was poor compared to the men they were replacing. The eighteen-year-olds had been living in a country that had endured four years of war. Everywhere the evidence of war weariness was obvious. Instead of the "liberation" that so many young men felt in joining up in 1914, the young men of 1918 had lived with four years of casualty figures in newspapers, of food rationing, and of restrictions of all kinds. They had lived a roller coaster of victories and defeats—the elation of the first day of the Somme followed hard by the shattering news of nearly 60,000 casualties. The emotions had been stirred by the first great tank victory at Cambrai only to be dampened when the German counterattack regained all the land lost. It was hard to expect these

[77]B. Pitt, *1918: The Last Act,* pp. 107–8.
[78]W. Moore, *See How They Ran,* pp. 167–68.

young civilians to possess the *esprit de corps* of the "Old Contemptibles" of 1914, who had added another glorious chapter to the history of British infantry. Those men had been professionals, hard-bitten craftsmen practicing their trade. They could fire fifteen aimed shots every minute and hit the bull's-eye every time. So shattering had been their fire at Mons and Le Cateau that the Germans had mistaken it for massed machine gun fire. Could these "babies" fight with the same spirit as their fathers? Only time would tell.

The German generals were drawing breath after the tremendous success of the 21 March offensive. Ludendorff had, in fact, already decided on a second assault, this time directed at the weakest point in the allied line, held by two Portuguese divisions. The Allied commanders, Pétain and Haig, were simply asking for trouble using such feeble units in the battle line, but so desperate was the manpower situation that they felt they had no real alternative. The British, displaying not for the first time in the war their irritating tendency to make fun of their friends, regarded their Portuguese allies as something of a joke. Even British generals called the Portuguese commander "General Bumface" as he did not speak English and was otherwise old and apparently incompetent. But the British were being unfair about him and his troops. The Portuguese troops were merely in France as a political gesture and felt no commitment to the struggle taking place. Whether the British continued to rule the seas and the sun never set on the British Empire were not major concerns to the poor conscripts. Whether Germany gained her "place in the sun" or the French their "blue line of the Vosges" mattered less than whether the Portuguese soldiers got something to eat and some warm clothes to wear in the unhealthy climate of northern Europe.

Just as the Allies were considering replacing the Portuguese with British troops the matter was taken out of their hands by the passage of events. The Germans had decided that the Portuguese offered little more opposition to their forward momentum than a house of straw in a wind tunnel. First the Germans flooded the forward British—and Portuguese—positions with mustard gas, followed the next day by one of Bruchmüller's barrages, as violent and as sudden as an electric storm. While the divisions flanking the Portuguese hung on, their southern allies saw little reason to lay down their lives in a cause for which they cared little. Many of the Portuguese surrendered to the Germans even before Bruchmüller's barrage, on the sensible, if unheroic grounds, that discretion was the better part of valor. Those who feared to advance toward the Germans—if only to surrender—threw off their boots and fled at top speed to the rear. This was riskier, but at least one preserved one's freedom of action. As we have seen so often, generalizations in warfare are even less rewarding than in other forms of human activity. Some Portuguese units, either because of strong leadership, fear of punishment, or bloody-mindedness, literally "stuck to their guns" and fought the Germans until they were overwhelmed. One unit fought to the last bullet and then bayonet-charged their attackers. On the other hand, those fleeing underwent a dual danger. Fearing the German shells behind them they ran toward advancing British troops, who fired at them as deserters. One British officer ordered machine guns to be set up, specifically to shoot at the fleeing Portuguese. From whatever cause, within a matter of hours the 2nd Portuguese Division had been entirely wiped out. Bruchmüller, meanwhile, was having a field day. Not content with concentrating his barrage merely on frontline positions, he had switched to hitting rear areas, notably the assembly points in villages behind the lines that had thought they were immune from fire. This was designed to spread the panic ever deeper into the British positions.

But as the British were pressed back resistance began to stiffen. Where soldiers would not stand and fight willingly, some of the tougher breed of officer were offering them the stark choice: fight or die. Lieutenant Colonel Seton Hutchison, leading a machine gun battalion of the 33rd Division, was swept off the road by hundreds of retreating soldiers, claiming to have mysterious orders to retire. Hutchison was unimpressed. He commandeered an ambulance to take him back to divisional headquarters, gained permission to move his machine guns forward, met opposition from a transport office when trying to requisition some lorries, and persuaded the man to see sense by knocking him unconscious with his revolver. He then raced back to the crowd of panicking deserters and stopped them in their tracks by setting up eight machine guns across their path. Any man who advanced a further step was immediately shot down and any officer who hesitated to do his duty was relieved of his command and arrested. Elsewhere, Brigadier General Cozier was acting with equal ruthlessness. He it was who mowed down the fleeing Portuguese and later wrote of his experiences in a book entitled *Men I Killed*, the men in question being fleeing British soldiers rather than Germans. Even Ludendorff himself later recognized the effectiveness of the British methods in regaining control of a panicking army. Anything less decisive, he observed, and the war would have been won by the Germans.

As the German advance continued, the morale of the two armies underwent a curious reversal. Success bred discontent in the German ranks, while disaster seemed, not for the first time, to steel the British resolve. Certainly British morale was already improving before Haig issued his famous order of 11 April: "There is no course open to us but to fight it out! Every position must be held to the last man; there must be no retirement. With our backs to the wall, and believing in the justice of our cause each one of us must fight to the end." One Royal Fusilier summed up the thoughts of many when he quipped, "What ruddy wall?" To many British soldiers the order was simply an insult. It sounded as if the High Command was panicking. They, the soldiers, were not. One senior officer refused to publish the order on the grounds that it would lower morale. Oddly enough, the Australians—usually the least sentimental of soldiers—found Haig's message inspiring. C.E.W. Bean, the Official Australian Historian recorded the Australian reaction: "The issue of this appeal has been criticized as unnecessarily alarming. Among the Australians, however, it had precisely the result intended—that of stinging them to the highest pitch of determination."

If the Australians were fighting at the highest pitch of their determination, the Germans were showing hardly any less themselves. On 14 April the Australians were assaulted by "miles of infantry" advancing against them in a full frontal attack. It was nothing less than murder. The Australians held their fire until they had unmissable targets and then massed artillery simply mowed down the Germans at a range of less than a hundred yards. One German officer displayed unbelievable courage. Shot down twice he still staggered to his feet bellowing *"Vorwarts! Vorwarts!"* before he fell finally, riddled with bullets. Four Australian machine guns spent hours killing thousands of Germans who marched toward them like moths toward a lamp. The Germans simply lacked the mobility to exploit their breakthrough. With few tanks and having lost aerial superiority to the Royal Flying Corps, the German offensive for all its massive potential was petering out in the same way that every offensive since 1914 had—in bloody failure. Now they were measuring progress in hundreds of yards and not miles, while their losses in experienced troops could not be made good. Ludendorff's great gamble had failed. While the British had suffered over 300,000 casualties in less than a month, German losses had been as great if not greater. Moreover, although the offensive had begun with a panicky flight by the British front line, it ended with 55 British divisions eventually repelling 109 German divisions. Individual British divisions took a perverse pride in their losses. In this competitive affliction, the 50th Division just pipped the 8th Division, suffering over 17,000 casualties against 300 less by the 8th.

In the words of William Moore, the battles "were devastating, but both sides learned valuable lessons. For the British, the most important of these was that to retreat does not mean to be beaten; for the Germans, that to advance does not mean victory." Even at the lowest point of their fortunes, with their battle line broken and panic afflicting the troops, the average British soldier remained stubborn and phlegmatic. Things might be bad, but their pride in their race and their history enabled them to maintain their fighting spirit. And when their own commander-in-chief called on them to fight to the end, with their backs to the wall, they felt not inspired but angry. They fought the

Germans to a standstill not because they were ordered to by their generals but because it was time to make a stand and put the Germans back in their place. The retreat had gone on long enough and the Huns needed a lesson. On such prejudices fighting spirit can be built.

11

The "Leathernecks" at Belleau Wood, 1918

The hostility shown by British and French soldiers to American troops when they first arrived in France in 1918 may have seemed inexplicable to the new arrivals. After all, were they not coming to save these very soldiers from the Germans? It was this sort of attitude that got the Anglo-American-French alliance off to a bad start. The Americans, in turn, found the British unfriendly and the French filthy. It is easy for later historians to understand the way the national characteristics might have grated together. The Anglo-French armies had been fighting the bitterest war known up to that time and they can hardly be blamed for thinking the Americans were making light of their struggle and their achievement. Certainly the Americans had the gaucherie of youth and inexperience. They often seemed loud-mouthed and arrogant, with their comments like: "Tell those Jerries we're here" and "Tell the Kaiser Bernie Bronowski's here." It was just a cultural clash; their lively sense of humor seemed at odds with the mood of Europe in 1918. Yet they had come to fight and, as one observer wrote, "These young men with their lithe and muscled bodies, their smooth faces and springy steps, resembled players on the gridiron but did not in the least evoke our heavy, unkempt and untidy poilus [French infantryman]." General Foch was pleased with what he saw. "We were impressed by the height of the men, by their well-fitting uniforms, by their physical development and poise, by their splendid health and vigor. If their gait lacked something of suppleness, this defect was compensated by the accuracy and precision of movement altogether remarkable." One might have thought the French were looking to buy the Americans as "thoroughbreds." Yet what was most appealing to the weary Europeans was the apparent sincerity of these young men, their lack of cynicism and their "quasi-religious seriousness." The Americans really believed in phrases like "La Fayette, we are here!", which they uttered to emphasize traditional Americo-French links. The French liked it; it was so deliciously artless. But the Americans soon began finding faults with their old allies: the French were really so *dirty*. The French protested that the Americans were always washing and wasting water. *Vive la difference!* The American commander, General Pershing, knew how to make enemies of his friends. As he wrote: "The morale of the Allies is low and association with them has a bad effect on our men. The fact is that our officers and men are far superior to the tired Europeans." By "association" Pershing meant "training." The French tried to pass on their experience from four years of war; the Americans did not want to listen. When shown gas masks, one Doughboy remarked, "I'm not going to wear one of those Goddamned masks for any lousy Hun!" They thought they knew best and were looking for a chance to prove it. Belleau Wood gave them that chance.

One young marine, Lieutenant Clifton Cates, wrote his mother an account of his journey to the front:

> We rode day and all night—it was an awful cold and dirty trip. If you can imagine about one thousand trucks lined up one behind the other and running as fast as possible. We passed through the outskirts of gay Paree and on thru numerous towns . . . [I] saw refugees plodding their way back and old men, women, and children: some walking and others on carts trying to carry their valuables back—it was the most pitiful sight I have ever seen and there is not a man in our bunch that didn't grit his teeth and say *"Vive la France."* "Do or die" is our motto—and the mother that can furnish a boy should say—"America—here's my boy, God grant that he may come back, if not, I am glad he died for a noble cause, and I am willing to give him to you."[79]

[79]J. Toland, *No Man's land,* p. 278.

The naïveté of these sentiments must have seemed refreshing—if rather quaint—to America's French and British allies. Few French mothers would have willingly handed over their sons as they might have done in 1914. The American approach to war seemed redolent of a simpler, nobler age; one that had been spared the heartbreak of Verdun or the Somme. The French people were amused by these tall, gangly young men who had come to save them. They looked so healthy when compared with the filthy, gnarled veterans who frequented the forward areas of both the British and French armies. The Americans looked as though they were going to a party. They had the confidence that comes with inexperience. Jean de Pierrefeu commented that, "New life had come to bring a fresh, surging vigor to the body of France, bled almost to death." But the Americans were unaware that her allies viewed her as a source of new blood—not a transfusion—but a sacrifice. Her young men came to bleed for France.

The US 2nd Division had already seen some fighting on the Marne and was eager to go over to the offensive. They were to receive their wish. On 6 June, 1918, the 4th Brigade of the United States Marines carried out one of the most heroic actions of World War I, as well as one of the most costly. It seemed as if the new American troops were determined to replicate the mistakes the French had made in 1914 in the battle of the Frontiers and the British a year later at Loos. They threw their bodies against the German machine gunners and paid for it in unnecessarily high casualties.

Belleau Wood was situated about eight miles to the west of Château Thierry, and was on a hill which dominated the countryside for quite some distance. As a wood it was very densely planted with trees, so much so that in some places they reduced visibility to just a few paces. The Germans had fortified the wood with machine guns, manned by veterans from the Eastern and Italian fronts. It was no place for the fainthearted. It was a natural killing ground, where the firing arc of the machine gunners interlocked.

At dawn on 6 June, the Franco-American artillery opened a bombardment on the German positions. Unfortunately, it was miscalculated and fired high over the wood and the village, doing no harm to anyone. After an hour of this thoroughly useless barrage, the marines climbed out of their trenches, totally bereft of artillery support. They advanced in four lines, separated by just twenty yards. Walking slowly toward the German positions they presented an unmissable target for the German machine gunners. Captain John Thomason records the casual approach of the Americans to the enemy fire:

> The platoons came out of the woods as dawn was getting grey. The light was strong when they advanced into the open wheat, now all starred with dewy poppies, red as blood. To the east the sun appeared, immensely red and round, a handbreadth above the horizon; a German shell burst black across the face of it, just to the left of the line. Men turned their heads to see, and many there looked no more upon the sun forever. "Boys, it's a fine, clear mornin'! Guess we get chow after we done molestin' these here Heinies, hey?"—One old non-com—was it Jerry Finnegan of the 49th?—had out a can of salmon, hoarded somehow against hard times. He haggled it open with his bayonet, and went forward so, eating chunks of goldfish from the point of that wicked knife. "Finnegan"—his platoon commander, a young gentleman inclined to peevishness before he'd had his morning coffee, was annoyed—"when you are quite through with your refreshments, you can—damn well fix that bayonet and get on with the war!"[80]

In terms of naïveté, nothing like it had been seen for four years on the Western Front and the Germans seemed almost too amazed to fire, but they soon recovered their equilibrium and began cutting swaths through the American ranks. Gaps appeared everywhere and the front line began to waver in the face of the storm of bullets. However, whistles blew and the marines, showing incredible bravery, faced to the front and kept advancing. Captain George Hamilton relates his impression: "Farther on we came to

---

[80]G. Chapman, *Vain Glory*, p. 618.

an open field—a wheatfield full of poppies—and here we caught hell."

But losses were so heavy that individuals began doubting their own training. Was this really the best way to combat machine guns? Abandoning the faulty doctrine they had been taught, the left wing—still two hundred yards from the German lines—dived full-length into the wheatfield for cover. In the center and on the right the marines were much closer to the German lines. Here some of them burst into a storming rush and covered the remaining distance in a few seconds. In Hamilton's words, "It was a case of every man for himself." Captain John Thomason of the 5th Battalion recalled what happened:

> Meanwhile, to the left a little group of men lay in the wheat under the very muzzle of a gun that clipped the stalks around their ears and riddled their combat packs—firing high by a matter of inches and the mercy of God. A man can stand just so much of that. Life presently ceases to be desirable; the only desirable thing to do is to kill that gunner, kill him with your hands! One of them, a corporal named Geer, said, "By God, let's get him!" And they got him. One fellow seized the spitting muzzle and up-ended it on the gunner; he lost a hand in the matter. Bayonets flashed in, and a rifle-butt rose and fell.[81]

Near the village of Bouresches, the marines faced the problem that the German machine gunners had taken up positions in the stone cottages. Under the only surviving officer, Lieutenant Robertson, the marines crept forward into the outlying houses and then rushed into the rest of the village, clearing the cottages at bayonet point. It took a long time to complete the operation, yet by 2:00 A.M. the village was in the hands of the marines, though just twenty men of the battalion that had begun the attack were still on their feet at the end.

In the center and on the right of Belleau Wood, things were looking bad. In many places the marines were left

[81]G. Chapman, *Vain Glory*, p. 621.

unsupported in the middle of the wheatfield, though some of them had reached the edge of the wood. One group had responded to legendary Gunnery Sergeant Daly's shout of encouragement, "Come on, you sons of bitches! Do you want to live forever." But still the volleys of machine gun fire ripped the wheat away from just above their heads. From the edge of the wood Sergeant Mervin Silverthorn saw the amazing sight of Major Sibley's battalion advancing in perfect order through the wheatfield and being pelted with bullets. It was like a scene from an epic of eighteenth century warfare. The men's courage was above praise; their leadership beneath it. From another part of the field Colonel Catlin referred to Sibley's advance as "one of the most beautiful sights I have ever witnessed." There was no yelling or rushing, just a deliberate march. It was magnificent but it was not war.

The fighting went on all day, and the marines trapped in the wheatfield were forced to spend the night there without being able to move until relieved in the morning. Those who did try to advance were caught in interlocking machine gun fire. The marines had managed to advance three hundred yards through the field into the storm of machine gun fire but no farther. The battle of Belleau Wood is often regarded as a one-day affair, though it dragged on for the rest of June. Yet it was the first advance of the marines that captured all the headlines. In the United States the press seized on this partial victory as if it came close to ending the war. In Chicago, the *Daily Tribune* reported: "Marines win hot battle; sweep enemy from heights near Thierry." The marines might have been relieved to hear that they had won; the men in the wheatfield may have wondered what a defeat felt like if this was victory. Nevertheless, the Americans were looking for a victory to establish their credentials as an army. As one reporter noted, "Everywhere one went in the cars, on the streets, in hotel or skyscrapers, the one topic was the Marines who are fighting with such glorious success in France." The fighting spirit of American troops in France—most of them thoroughly sick and tired of hearing about Leathernecks and determined to show that they could do just as well—rose appreciably during this well-publicized event.

After three days' fighting—on 9 June—the marines were withdrawn from the wood, which was then blasted by two hundred Franco-American guns. Unfortunately, the barrage flattened many trees and made the wood even more of an obstacle to attacking troops. When the marines attacked again they found the same problems as before. Nevertheless, showing incredible fortitude, the marines crawled through the mass of undergrowth and fallen trees. Fighting with cudgels and grenades in the chaotic conditions, they gained a foothold in the wood on the night of 9 June. However, even now the Germans would not give ground. The marines were withdrawn again and a further barrage of the wood took place. On 12 June, the marines tried again and this time were successful in clearing a large part of the wood. However, it was not until 25 June that the whole of the wood fell into American hands. American losses had been very heavy. The 4th Marines Brigade which had begun the attack lost 5,711 men, including half of all officers. In the context of the war as a whole, the struggle for Belleau Wood might seem insignificant, but its importance for the newly arrived American troops was inestimable. Any sense of inferiority they may secretly have felt toward the battle-hardened British or French veterans disappeared like a morning mist in the heat of the sun. The marines had lived up to their reputation as the toughest of fighters. The Germans had come to regard the Americans as wild men, fearless in the attack and tenacious in defense. Belleau Wood established a legend that would fortify many young marines in their battles ahead, notably in the Pacific in the Second World War.

12

"Run, Run, the Bogeyman Is Coming!"— The Fall of Anual, 1921

The nineteenth century was the highpoint of European colonization, notably in Africa, where the soldiers of most of the European powers found themselves involved in wars against the indigenous peoples. Although the Europeans suffered some startling reverses against the natives, much of the time their soldiers were involved in long periods of garrison duty, far from home and the pleasures of civilization. In such circumstances morale was difficult to maintain and fighting spirit suffered, notably in the case of armies where discipline had been allowed to relax. The consequences of this decline of morale could be quite dramatic and few examples were more remarkable than that which befell the Spanish army in Morocco in 1921.

The Spanish government had adopted a policy of administering Morocco by means of sprinkling small garrisons of troops across the country, yet ruling centrally from the city of Melilla. This plan, of spreading hundreds of small blockhouses and forts—often with no more than twenty troops in each—across the vast areas of Moroccan desert, was particularly ill-advised in terms of the effect it had on the morale of the troops. The sense of isolation was immense and frequently overwhelming. The Spanish soldiers were poorly fed—on a diet of just coffee, beans, rice, and bread—and endured living conditions actually worse than the tribesmen over whom they were supposed to be maintaining order. Their pay was so low that it was less than a third of what the Berber tribesmen could earn for laboring on the roads. The Spanish troops possessed little self-respect, being conscripted mainly from ill-educated or even illiterate peasants in Spain, who had little experience of using modern machinery and found even their own weapons difficult to use effectively. When they were equipped with up-to-date rifles, they sold them to the Rif tribesmen—their most dangerous adversaries—to get more money to buy food, cigarettes, and wine. The sale of their modern weapons meant that many Spanish conscripts were forced to use ancient firearms, some of which were so poor that they could fire a bullet no farther than a hundred yards.

Under these circumstances, only the most active leadership could have restored the morale of the men and prevented a catastrophe. Unfortunately, the Spanish officers took their service in Morocco far too lightly. Junior officers even resorted to stealing from army stores to improve their own living conditions, while their seniors spent much of their time on leave in big cities like Melilla or even in Spain itself. The idea of visiting the outlying garrisons was never seriously considered, and no officer considered his men as anything other than the lowest ragamuffins. The Spanish recruits responded to such leadership by attempting to avoid as much work as possible. Malingering became the main occupation, with self-inflicted wounds being a common alternative to duty. Venereal diseases were willingly contracted by the Spanish recruits as a way of earning a return to Spain, while tobacco was chewed to simulate the effects of jaundice, nettles were placed on wounds to make them fester, and heated coins applied to the skin to make sores. How effective this deception was depended on the size of the garrison, for few of the smaller forts had doctors or any form of medical service.

Like the French army in 1940, the Spanish colonial army in Morocco was ripe for disaster in 1921. The Beni Urriagali tribe, under Abd el Krim, aimed to prevent Spain expanding her territory into the Rif. Krim was aware of the poor quality of the Spanish troops and knew that if he could raise the tribes against their masters, Spanish rule could be overthrown. In June 1921, the Spanish commander, General Silvestre, marched into Abd el Krim's territory and Krim raised the banner of revolt by attacking the Spanish fortress of Anual. Immediately it became obvious that the Spanish troops could not face the fierce tribesmen, particularly as the latter were equipped with the modern rifles that the Spaniards themselves had sold them. As the Rif tribesmen besieged the large fortress-town of Anual, Silvestre simply told his officers to "pull out by surprise." With orders like this from the commander, the common soldiers cracked. The gates of the fortress were thrown open and hundreds of Spanish troops began a panicky retreat. Silvestre simply shouted after them, "Run, run, the bogeyman is coming." Some were lucky and escaped in cars or lorries, but others tried to outrun the Rif on horses, or even pathetically on foot. Soon the desert was heaped with the bodies of the Spanish soldiers who had lost all fighting spirit and were cut down by the tribesmen as if they had lost all power of resistance.

After the capture of Anual, the other Spanish garrisons fell like a series of dominoes. Even previously passive tribes took up arms to join in the massacre of the Spanish army. Lacking properly fortified positions, no garrison was able to hold out for long. In the next few days, 19,000 Spanish soldiers were butchered in the desert—most by sword or knife—by Krim's army of fewer than 3,000 warriors. Senior officers in Melilla had shown cowardice by refusing to lead a relief force to help Silvestre's men and, instead, hiding in cellars or choosing the very moment of battle to take leave in Spain. Stories spread of officers fleeing from their garrisons by taking the only car and leaving their men to the mercy of the Rifs. The famous French colonial soldier and administrator, Marshal Lyautey, remarked on hearing of the disaster at Anual, "The Spanish soldier, who is as brave as he is long-suffering, can, under another command, know better days."

13

"A Nation in Arms"— The Fall of France, 1940

If ever an army was a mirror of the nation, it was that of France in 1939. Superbly equipped and, contrary to popular belief, with more and better tanks than the Germans, it boasted the strongest defensive line ever built. Yet even such offensive and defensive power was insufficient to inspire the French soldier with self-belief. Twice before—in 1870 and 1914—the Germans had invaded French soil. Would the mighty Maginot Line be enough to keep them at bay a third time? The French were unsure: alternately complacent behind their wall and yet defeatist, their confidence in their army was at a low ebb, and their confidence in their British allies was even lower. Many Frenchmen seemed to hate and fear the British almost more than they did the Germans.

Adolf Hitler had already evaluated the fighting spirit of the French army as very low. As he told the German generals, "The French people think only of peace and good living, and they are torn apart in Parliamentary strife. Accordingly, the army, however brave and well-trained its officer corps may be, does not show the combat determination expected of it. After the first setbacks, it will crack." Hitler was right. Morale among France's soldiers was at rock bottom during the fall of 1939, in the period that was known as the "Phony War." At a time when the Germans, fresh from their triumph over Poland, were preparing for the war in the west, the French were turning their swords not into plowshares but into bottle openers and their bayonets into corkscrews. Drunkenness was rife among the army of the Third Republic; too many soldiers were seeking salvation from their fears in alcohol.

The morale of any army suffers when the common soldiers feel that they are being treated less well than their officers. Compared with the far more democratic Germans—democratic in the sense that the officers roughed it with their men—the French soldiers were exposed to harsh conditions at the front. These conditions were not the product of wartime necessity but of simple neglect. Shortages of boots and blankets were unforgivable and along the Sedan front—where the wretched 55th Division held the line—the soldiers were forced through lack of billets to sleep in the stables with their horses. An American observer, who had been a nurse with the French army in 1917, noticed how different things were in 1939. She noted the absurd contrast between the officers who had "well-polished nails and brilliantined hair" and the ordinary soldiers "in their grubby barracks and the dreary canteens in which they could spend their fifty centimes a day." The French soldier had been paid the same in the First World War, which had been the cause of much discontent when he realized how much more his British and American allies were receiving. One result, in 1939, of paying frontline soldiers such a pittance was that most of them used their ten days' leave doing other jobs, like driving Paris taxis. On the other hand, officers were frequently able to maintain their civilian businesses even when in the front line, running their own private cars and keeping their wives and mistresses in lodgings within officially designated "Zones of Armies," from which civilians were officially prohibited. It was a formula for slackness and inefficiency.

For the French, in their massive fortifications, the greatest threat to fighting spirit and military efficiency was boredom. It is impossible to maintain peak combat performance when soldiers have too much time to think about the home comforts that they are missing. The ironically termed "Sitzkrieg"—during which the French and German troops sat and watched each other in the early months of the war—undermined warlike intentions on the French side and fed the boredom to which they so easily succumbed. For the

Germans, however, it was just the calm before the storm that they would release in May 1940. Service in the Maginot Line, during the harsh winter of 1939–40, was described by one French soldier: "Before you an unknown countryside, a black night. The nearest post was several hundred yards away. Your feet are frozen in their stiff boots. Your helmet weighs heavily. Your eyes are tired from looking without seeing." It is hardly the response of a soldier with high morale. Such guard duty may have been the lot of soldiers throughout history, but too much "passive" service of this kind can blunt fighting spirit and unless this is carefully nurtured, the defensive mentality, which service behind fortifications can foster, will erode the soldiers' capacity to respond to an "active" opponent. This is exactly what happened to millions of French soldiers in May and June 1940. As one officer pointed out, "We are no longer fighting the Germans we are fighting *ennui*."

On the other hand, the British troops in France at that time, members of the BEF, were subjected to a regime of apparently mindless activity. While the French leaned on their rifles, dozed at their posts, and dreamed of better days, the British soldiers were ordered to dig holes and then refill them, paint coal white, and build and then take down useless structures. The point was simple. The British kept busy, fit, and alert. They may even had grown angry at the futility of what they were doing. But, above all, they remained hard and battle-fit. When British general Alan Brooke visited French Ninth Army headquarters he was shocked by what he found:

> I can still see those troops now. Seldom have I seen anything more slovenly and badly turned out. Men unshaven, horses ungroomed, clothes and saddlery that did not fit, vehicles dirty and complete lack of pride in themselves or their units. What shook me most, however, was the look in the men's faces, disgruntled and insubordinate looks and, although ordered to give "Eyes left!" hardly a man bothered to do so.[82]

[82]A. Horne, *To Lose a Battle,* p. 222.

Discipline in the French Army in 1940 hit rock bottom. It seemed that nobody could really be bothered to fight a war. One officer observed a soldier, who had been ordered to place demolition charges under one of the Meuse bridges, abandon his post, put down his rifle, and join one of his friends, who was fishing some distance away. To make matters worse, the French police reported that soldiers were looting and vandalizing the evacuated areas they occupied, as if they were a conquering army. Some looters and pillagers had even been shot. Much of the trouble stemmed from the response the authorities adopted to the problem of the soldiers' boredom. While the British believed in keeping everyone busy working hard, the French decided that the only answer was to extend the licensing hours so that the men could drink at virtually any time of day or night. The consequent rash of drunkenness contributed to the overall impression that the French army had neither the ability nor the inclination to face the Germans.

And so developed the idea of "French Leave." At first surreptitiously, and then more openly, the frontline soldiers began drifting home for long weekends, often being absent without permission for days at a time. Rather than stamping on this creeping malaise, the High Command wrung its hands in despair. One First World War hero reflected after visiting the front, "I seemed to encounter slackened resolve, relaxed discipline. There one no longer breathed the pure and enlivening air of the trenches of 1914–18." The men at the front would have tapped their foreheads in despair at this. They would have wondered how the old man could have breathed "pure and enlivening air" in the trenches at Verdun in 1916. Their bitter cynicism would have asked him what he had been fighting for a generation ago: just so that they could stand in the same line against the same enemy less than a quarter of a century later? It was this that was at the root of the French problem. Themes and motives that had sent men into war and apparently willing sacrifice in 1914 were no longer apparent. What was this new war for? If they were fighting to save Poland, they were already too late. Poland had fallen and France had not lifted an active finger to help her. Why go on with the farce?

Faced with the collapse of morale in the Army, the French High Command simply hid its head in the sand. Even the postal service, recognized as vital to army morale in wars of the twentieth century, was allowed to molder. Men were often left for six weeks at a time without news from home, and this merely increased the pressure of soldiers to slip away home when the urge to do so proved irresistible. The Germans, of course, were perfectly aware of the disintegration of French morale and did everything they could to exploit it. Two areas were ripe for exploitation: French dislike of the war and, above all, French dislike of their British allies. Anglophobia was almost France's favorite pastime, and Joseph Goebbels soon set about exploiting it. The Luftwaffe dropped cartoons on the French lines. One, named "The bloodbath" showed a little Frenchman and a large pipe-smoking British "Tommy" standing on the edge of a blood-filled pool. In the second picture they both prepare to jump. In the third picture it is seen that only the French soldier has jumped into the pool. The fourth picture shows the French soldier up to his neck in the blood, while the British "Tommy" is walking away laughing. A second cartoon showed British officers in Paris fondling a half-naked woman, while a French soldier was shown keeping guard on the Maginot Line. Such crude propaganda could never have worked on troops with high morale, who would have seen them for what they were. But on troops already deeply depressed by their situation, they played on nagging doubts. Why, for instance, had Britain, with a much larger population than France, sent so few soldiers to hold back the Germans? Was there truth in the German slogan: "*Les Anglais donnent leurs machines, les Francais donnent leurs poitrines*"? Many Frenchmen believed it had been like that in 1914 as well. As usual, it seemed, the British would fight to the last French life. During the winter of 1940 a common phrase among the ordinary French people was, "Have you seen the English?"

General Heinz Guderian's breakthrough at Sedan on 13 May, 1940, was one of the turning points in modern history. Cracked forever was the image of France as a great military power. Seventy years of Franco-German conflict for hegemony on the European continent was settled. In spite of the great struggle of 1914–18, the conflict was finally resolved when the ultramodern armored divisions of Guderian fell upon the second-rate 55th Division of General Lafontaine. The 55th was a ridiculously inadequate opponent for Guderian, made up as it was of elderly, unfit reservists. Even later French apologists have admitted that resistance was "feeble." Although the collapse of a division should not have been the end of resistance at a stroke, the consequences of Guderian's success for the French were astonishing. The German *blitzkrieg* simply shattered French morale and led to a virtual collapse of combat effectiveness in large parts of the French Army. Hysteria on a massive scale was the order of the day. Yet, as Alistair Horne has pointed out in his book *To Lose a Battle,* the Poles had faced the terrifying Stuka dive-bombers and the Panzer divisions with far more courage and spirit than the French, and without anything like the material strength the French possessed. For Horne, the answer was that the Poles were fighting for their lives, and knew it. The French soldiers did not know what they were fighting for. Few of them expected to see their nation disappear under a tyrant's heel as did the Poles, and for them self-preservation was still the overriding priority. There was little of the self-sacrifice that had marked the fighting in 1916 for Verdun, a national symbol with which all Frenchmen could identify. As Hitler had so astutely explained to his generals: France was riven with defeatism and, with her army mirroring the state of the nation, it would only take a single breakthrough for the whole military facade that was France in 1940 to come crashing down.

The French generals were blind to the weakness of morale in their own troops. At the very moment that his division had cracked at Sedan, General Lafontaine was poring over his maps, planning moves by imaginary regiments, seemingly oblivious to the fact that a tidal wave of panicky soldiery was about to sweep him away. One of his officers suddenly observed:

> A wave of terrified fugitives, gunners and infantry, in transport, on foot, many without arms but dragging their kitbags, swept down the Bulson road. "The tanks are at Bulson!" they cried. Some were firing their rifles like madmen. General Lafontaine and his officers ran in front of them, tried to reason with them, made them put their lorries across the road . . . Officers were among the deserters. Gunners, especially from the corps's heavy artillery, and infantry soldiers from the 55th division, were mixed together, terror-stricken and in the grip of mass hysteria. All these men claimed actually to have seen tanks at Bulson and Chaumont! Much worse, commanders at all levels pretended having received orders to withdraw, but were quite unable to show them or even to say exactly where the orders had come from. Panic brooked no delay; command posts emptied like magic.[83]

Once begun the panic was almost impossible to stop. In one case, ten gunners and a medical officer fled from Sedan in a lorry. They were finally stopped six days later in Auxerre, hundreds of miles from the front. When the medical officer was questioned about his actions, he explained, "But, *mon capitaine,* bombs were falling . . ." Meanwhile, General Lafontaine was still at his headquarters trying to redress the faults of a whole generation in a few hours. Instead, as General Ruby relates, the tide of panic swept past him as if he were some outcrop of rock, immovable yet without understanding. "The flood of fugitives traverses the villages without pause; all the echelons of the division accumulated in this region—fighting units, regimental H.Q.s, supply columns, vehicle parks—all are heading for the south, swelled by stragglers; as if by magic, their officers have naturally received a mysterious order to withdraw. Barriers established by the military police are swept aside."

The panic spread as news of the disaster that had overtaken the 55th Division became generally known. The cry was heard everywhere that the German tanks were in the next town. As General Menu remarked, "Riflemen and

[83]A. Horne, *To Lose a Battle,* p. 348.

machine-gunners got up and fled, taking with them in their flight those of the artillerymen who had not already beaten them to it, mingled with elements flooding backwards from the neighbouring sector." Even as the French commander-in-chief, General Maurice Gamelin, tried to order reserves forward, they found the areas just behind the front line to have been devastated by their own retreating men. Bridges and telephone exchanges had been destroyed on "superior orders," though nobody was ever able to discover who had given them. A French tank commander described the chaos he encountered:

> One squadron of the 5th Cuirassiers discovered a sergeant-major who had torn off his badges of rank, so as to take flight more easily, both from the enemy and from his own responsibilities; later a lieutenant of the divisional anti-tank battery reported fugitives among whom were found two lieutenants who had similarly degraded themselves; he made this troop turn about under threat of his sub-machine gun. All these cowards were causing terrifying rumours to run around, notably about the aerial bombardments to which they had been subjected, and about the avalanche of tanks which was closely pursuing them . . . One had to seek proper justification for the rout . . . The fugitives said that they had been pursued by formidable masses of tanks (some spoke of 400, others of 500, or even 5000!).[84]

In the panic there were numerous "blue-on-blues" and near Raucourt there was a full-scale battle between two French battalions, with heavy casualties on both sides. Significantly, although much of the area south of Sedan was littered with abandoned French equipment, the fleeing troops had still found time to loot the villages and the medical stores looking for liquor.

Even now the collapse of fighting spirit should have been restricted to the two second-rate divisions that had broken—the 55th and the 71st—but soon it was spreading like wildfire through much of General Hutziger's Second Army

---

[84]A. Horne, *To Lose a Battle*, p. 384.

and General Corap's Ninth Army. The American correspondent Drew Middleton observed the ragged remains of the Ninth: "They were the clerks, the cooks, the anti-aircraft, and the heavy artillery of the Army that had been decisively beaten and routed. As they sat on their horse-drawn carts in their dirty uniforms, they did not look like soldiers but like gypsies." Of the Ninth's 70,000 men, fewer than one in ten still had his rifle, yet few of these refugees carried any sign of injury or wounds. Rumor alone, not the Germans, had caused an army to panic and flee.

The French collapse at Sedan had been a vast self-inflicted wound. How otherwise could one explain the extraordinary incident witnessed by British General Sir Edward Spears:

> A group of the best heavy guns in the French Army, the 155-millimetre Rimaillots, was halted near Laon when a pale-faced Staff Officer appeared declaring that he had come post-haste from Corps H.Q. to say that a German Panzer division was converging on them and would be there in a matter of minutes, and the Corps Commander adjured them as good Frenchmen not to allow their guns to fall into enemy hands. Within a few minutes 35 of these priceless guns had been damaged beyond repair. No such order had been sent from Corps H.Q.[85]

Meanwhile, the British Expeditionary Force, far to the north, was pulling back toward the Channel, astonished by the sudden collapse of what they had previously assumed to be their invincible French allies. Could these be the same troops, they reasoned, who had fought the Germans to a standstill in the great battles of 1914 and 1915? The British troops, far less well equipped than their French colleagues, nevertheless revealed a solid combat effectiveness when they fought the Germans. General Reinhardt, fresh from his easy triumph over the French, observed that his first encounter with the British, "in contrast with the French, [the British] caused surprise by their tough way of

[85]A. Horne, *To Lose a Battle,* p. 517.

fighting and are only overcome after a one-hour battle." In terms of fighting spirit one has only to think about the different national approaches to the "Phony War." The French soldiers drank and took unofficial leave; the British dug holes and filled them in. Certainly the British soldiers must have resented the sheer futility of what they were doing. Yet, it was hard work and it tired them out so that they slept well, untroubled by the boredom from which the French soldiers suffered so badly. Moreover, this extension of British "bull" maintained discipline at a time when it was crumbling in the French ranks. The tragedy was that the French only began to fight when it was too late. The French soldier had not known what he was fighting for in September 1939. As we have observed, the Poles fought with far greater determination than the French precisely because they knew what was at stake. The French, unaware of the sheer size of the disaster that was facing them, fought faintly in a poor cause. However, once defeat stared them in the face, and defeat moreover at the hands of their much-feared neighbors, some units began to offer stiff resistance to the Germans. But it was "too little, too late." Once the Maginot Line had been both breached and bypassed, no amount of heroic resistance would prove a substitute for a negative national policy of defense. The Maginot Line had made Frenchmen both complacent and ultradefensive. The Germans, on the other hand, had embraced the lessons of mobile warfare that their enemies had taught them in the last months of the First World War. Against the dive-bombers and tanks of German *blitzkrieg,* France had offered the fighting spirit of her soldiers. It was never going to be enough because Frenchmen did not believe in the war. Even worse they had ceased to believe in themselves.

14

The End of "Fortress Singapore," 1942

The collapse of "Fortress Singapore" in February 1942 was the greatest defeat ever suffered by British arms and the most humiliating episode in the whole history of the British Empire. Although it can be argued that the defeat was a political one—the result of strategic errors going back half a century—this cannot hide the fact that the Japanese conquest of Malaya that preceded it and the subsequent assault on the island of Singapore saw a dreadful collapse of military morale and a shameful loss of fighting spirit by the large garrison which, even as they surrendered, still outnumbered their opponents by nearly three to one. Few can look back on the events of 1942 with other than a sense of shame and for the Australians—those doughty fighters at Gallipoli and on the Western Front between 1916 and 1918—the story has proved too painful to contemplate. Accounts of Australia's part in the Second World War have tended to draw a veil over what happened in the last few weeks of the Singapore campaign. For many reasons morale collapsed, but as the fighting spirit of the troops disappeared it was replaced by such a disintegration of military discipline that the troops became almost more of a threat to the people they were defending than the enemy. The story is a pitiful one, but it shows how once self-discipline has gone men can succumb to their own worst fears and find no rest other than in mad flight.

At the outset it is only fair to say that the military garrison at Singapore had never been intended to stand alone. It had relied for its defense not only on its own guns and men but on the strength of the Royal Navy. In wartime it had been expected that any enemy invasion force would have been destroyed by British warships long before it reached the coast of Malaya or the island of Singapore. So strong indeed was British naval power after the First World War that it had been thought inconceivable that any Far Eastern power—and this could only mean Japan, with whom at that time Britain was friendly—could ever threaten Britain's imperial possessions. However, the rise of Japan militarily presented Britain with the problem of dividing her strength between the European and Far Eastern theaters. With Germany always the greater threat, in the interwar period, Britain found herself unable to commit ships permanently to the great naval base at Singapore. This was a source of grave worry to Britain's dominions of Australia and New Zealand, and so Britain made a promise that, even as she was making it, she knew she could not keep. Britain promised that in the event of a war with Japan she would guarantee to send her fleet to the Far East. This may have lulled the Australians into a false sense of security. Even by 1921, Japan was a formidable naval power and only the best of the British fleet would have proved good enough to match her. Unfortunately, with a hostile Germany threatening Britain in Europe during the 1930s, there was never any question of Britain being able to send her fleet eastward. Moreover, Britain could neither send her best troops to garrison Singapore, nor equip the colony with her best aircraft. Frankly, the British were simply crossing their fingers and hoping that Japan would not exploit Britain's European difficulties in 1939. In fact, everything was allowed to take priority over the defense of Singapore. When war broke out in Europe, the best Australian and Indian troops were sent to North Africa; the best available fighter planes were sent to help Russia, and the most modern British warships were kept in home waters to face the danger of German surface vessels. When war with Japan eventually came in 1941, Britain's bluff had been called. She no longer had the military strength to fight a global war against two first-rate military powers. The Australians felt that they had been betrayed: Britain could not afford to send a fleet to face the Japanese. Instead Prime Minister

Winston Churchill sent a symbolic offering—in some ways no more than a sacrifice to appease his critics. On the absurd assumption that the Japanese could be deterred by the mere presence of the Royal Navy in whatever strength, he sent out the new battleship *Prince of Wales* and the old battle cruiser *Repulse,* without any air cover. They were symbols—but symbols of Britain's weakness not of her strength. To offset the lack of proper air defenses in Malaya and Singapore, Britain increased her military garrison. However, the troops were essentially second-rate, consisting of untrained Australian and Indian units, while the best continued to be used against the Germans in North Africa and the Balkans. Churchill was taking a great risk. His two "battlewagons" were little more than a bluff to the Japanese, even though the ground troops in Malaya and Singapore still believed that the two ships could in some way deter the Japanese from invading Malaya. Rarely has so much ridden on the decks of just two ships. But the Japanese, who had not been deterred from attacking the entire American fleet at Pearl Harbor, were not going to be rocked on their heels by Britain's symbolic gestures. Shortly after the two British ships arrived off Malaya they were overwhelmed by Japanese torpedo-planes and sunk with terrible loss of life.

The Japanese invasion of Malaya and the subsequent loss of the *Prince of Wales* and the *Repulse* struck a devastating blow to British morale. Churchill's bluff had been called and it was shown to have been a very shallow one indeed. The effect on the ground troops in Malaya was very bad, for they had hoped that the two warships would be able to sever communications between the Japanese troops, if they landed in Malaya, and their High Command in Japan, perhaps even preventing supplies, ammunition, and reinforcements being brought in to aid them. Now that hope had gone their minds turned only to escape. They thought of the epic evacuations for which the Royal Navy was famous, like Narvik, Dunkirk, and Crete. Defeatism became widespread even though on the ground the British defenders heavily outnumbered the invaders. Preconceptions about Japanese equipment and the fighting quality of their soldiers were now proving to be terribly wrong. The British had underestimated their enemy, always a dangerous mistake in war. Historian Louis Allen records two examples of this:

> I was amused by one battalion commander, Brooke-Popham wrote to General Ismay, who while we were standing together looking at his men said, "Don't you think they are worthy of some better enemy than the Japanese?" . . . I also got a similar remark from the Colonel of the Argyll and Sutherland Highlanders yesterday; he had trained his battalion to a very high pitch for attacking in the type of country one gets near the coast and said to me, "I do hope, Sir, we are not getting too strong in Malaya, because if so the Japanese may never attempt a landing."[86]

Allen comments pointedly that the colonel who feared himself too strong was, within a year, to lose his entire brigade to a single Japanese tank column. A further shock to British prestige came in the realization that in their military technology the Japanese had advanced way beyond the Allies. It had been a popular belief that Japanese pilots and planes must inevitably be inferior to those of the Western powers. This was ethnocentrism perfectly expressed. When the Japanese A6M2 Zero-sen naval fighter appeared over Malaya and promptly cleared the skies of the Allied planes it aroused a feeling of psychological shock. Surely only white men could build and fly such a plane? The pilots must be Germans and the plans based on those of a Western power. Such attitudes were an attempt to bolster the belief in white supremacy and to discredit everything Japanese. However, even more than the United States, Britain refused to face the truth about the Japanese, and by doing so ensured the destruction of her own position in the Far East.

At the start of the war in the Pacific—in December, 1941—Britain had a total of 90,000 troops in Malaya and Singapore, of whom 20,000 were British, 15,000 Australian,

[86]G. Regan, *Someone Had Blundered,* p. 273.

37,000 Indian and 17,000 or so local Asian. Altogether, this was 40 percent under strength, or 17 battalions light. However, it was not simply a question of numbers. Malaya Command was something of a backwater and the troops there were poorly trained and had many second-rate officers. A particular problem was to be found in the Indian units, which comprised the largest single element in the command. A century or more of experience had shown that the key to the effectiveness of Indian units was a close relationship between the troops and their officers, many of them white, who were fluent in native dialects. At the outbreak of war in Europe in 1939 the best Indian troops were sent to North Africa and the Middle East. The result was that the less good units were stationed in Malaya and the Far East. Moreover, the talented white officers were not left in the backwater but were moved to active service or staff positions in the theaters of war, while their places were taken by inexperienced conscripted white junior officers, who were unfamiliar with the troops and their ways. The consequent decline in standards was inevitable. In fact, many of the Indian units were at best half-trained and had no experience at all in armored warfare (there was not a single tank in the whole of India!) The Australians were no better placed. Noel Barber records that,

> All except one battalion had been "made up" with many untrained reinforcements. Some had sailed from Australia within two weeks of enlistment. Many hardly knew how to handle a gun. The decision to select these untrained Australian reinforcements for Malaya was unfortunate.[87]

The failed policy of deterrence toward Japan had meant that troops found themselves in the front line against a well-equipped and professional enemy, whose capabilities had been seriously underestimated. The result was that the troops in Malaya lacked adequate training in antiaircraft defense and had no antitank experience. Rather than send-

[87]N. Barber, *Sinister Twilight,* p. 125.

ing antitank guns, London sent instead manuals on how to combat tanks without guns. Through an oversight on someone's part these were kept in an office cupboard and never issued until discovered by Brigadier Ivan Simpson. This able and energetic officer personally devised a system of opposing enemy armor specifically for the Indian troops, but was forbidden to circulate it by Malaya Command. Simpson's reports on the state of defenses in Malaya made depressing reading. He found no evidence of any coherent policy of defense for the whole of the Malayan Peninsula. When he suggested to the c-in-c, General Percival, that something should be hurriedly constructed, he was told that defensive positions would be bad for morale and would be an admission that some areas of Malaya could not be defended. This attitude is quite remarkable: the fact that the truth was unpleasant did not make it any less the truth.

Japanese infantry tactics came as a shock to the British. Their mobility—many of them using bicycles—made light of the difficulties the British had anticipated they would face in advancing through Malaya. At the head of their forces was a "shock-group" of light armor. Even though their tanks were relatively feeble in comparison with the best British, German, or American models, they were supremely effective because they were the only tanks in Malaya. Japanese tactics consisted of using the roads to advance until they made contact with British forces, whereupon they moved into the jungle at the side of the roads to infiltrate, outflank, and encircle their enemy. The British were tactically outclassed.

Just as the whole British army in Malaya was defending a naval base that was useless to Britain, so on a smaller scale her troops were defending airfields which were only as good as the planes that used them. The inferiority of British planes made it pointless to leave troops scattered throughout the countryside. The Jitra line, which was supposed to hold for three months, broke in fifteen hours because of feeble resistance. Clearly only brilliant generalship and an influx of experienced reinforcements could save the whole of the peninsula. Of brilliant generalship there was absolutely no sign, but London decided to send more troops as if—at this eleventh hour—it was really possible to turn back the tide. It was decided to send the 18th British Division and the 17th Indian Division, which had been earmarked for North Africa, to Malaya instead, as well as an armored brigade and eighteen air squadrons. It was remarkable how troops that had been unavailable when they could have been of some use were suddenly discovered in time to swell the number of prisoners of war. In fact, they only served to bring the garrison up to its full strength of forty-eight battalions. Had they been in Malaya at the time of the Japanese landings, they could have been used for counteroffensive operations, but by now it was far too late. In any case, the quality of the new troops left much to be desired. As a recent book suggests:

> An ugly rumour abounded that some of them [the new Australian troops] were the sweepings of the Sydney dock area and that others were convicts who had been paroled providing they "volunteered" . . . Whatever the truth may be, the men were definitely not professionally fit for combat and, what is more, they had already witnessed a major breakdown of discipline among their better trained comrades in Fremantle. It was these men, from both sections of the reinforcements, who were to be spread around the various Australian battalions in Singapore, thus diluting the combat efficiency, discipline and morale of all.[88]

As the Japanese advanced down the Malayan Peninsula, unit after unit of British, Australian, and Indian troops cracked and began retreating in panic. As they did so they drove before them huge numbers of refugees, fleeing from the fighting, and crossing the causeway into Singapore to swell the civilian population. The new troops arriving in Singapore had no chance of reversing the flow of battle. In the first place, they were not acclimatized; the 18th British division had been at sea for eleven weeks and had, in addition, been trained in mechanized warfare, which none of

[88]P. Elphick, *The Pregnable Fortress,* p. 305.

them were to experience in Singapore. Their arrival, as the last troops were pulled back across the causeway, was a strategic absurdity. They were leaving their ships in order to become Japanese prisoners of war and Churchill and his chiefs of staff must have known this. Britain was facing humiliation; yet was it more humiliating to lose Singapore without making an effort or to fight to the last man and still lose it? The answer belonged to the world of politics.

In January 1942 the British government was facing a certain disaster in Singapore. The decision to send substantial reinforcements to a garrison which could only hold out for weeks at the most was a serious mistake. Politicians, particularly from Australia, stressed the fact that to abandon Singapore was to renege on promises made in the interwar period. Trapped as he was, Churchill's response was to call on the garrison to "fight to the last man." These were brave words spoken in London, thousands of miles away. But in Singapore the garrison was not listening. They were trapped and they did not intend either fighting to the last man or waiting to become prisoners of the Japanese. What followed was mass hysteria on the part of the troops. Yet it was not madness—except in military terms—for many of the troops had a very clear idea of what they were going to do. They were not going to hold their lines. They were going to run back to Singapore, get drunk, and loot the place before the Japanese could. Then they were going to get aboard the boats that were leaving the harbor for Australia. And if anyone tried to stop them—officers, police, or even the Japanese—they would see how a trapped Aussie could fight. The extracts that follow are not intended as criticism of the Australian fighting soldier, whose achievements are too numerous to need to be restated, but the truth was that the Singapore garrison, particularly the Australian component, displayed some of the worst aspects of an army in panicky rout. A military intelligence officer from the Australian army witnessed part of the rout:

> Hundreds of bedraggled Aussies were streaming down Bukit Timah Road on the way to the city. The Military Police (UK and AIF) attempted to check them but they were in no

> mood for "homilies" from "Red Caps." Some paused long enough to accept a cigarette, light it and say, "Chum, to hell with Malaya and Singapore. Navy let us down, airforce let us down. If the *bungs* [natives] won't fight for their bloody country, why pick on me?"[89]

An Australian naval officer reported how one Australian unit had cracked and run:

> And the Australians were always the worst. The best when they were good, the worst when they were bad. There was one group of Australians that had been getting a terrible pasting and the Japanese were coming across every day and machine-gunning them from the air. One day they just threw down their weapons and said they were not going to fight any more . . . [their] senior officer almost pleaded with them but to no avail.[90]

The British were hardly appreciative of the Aussie collapse as this comment by Colonel Ashmore shows:

> The AIF had by this time (10 and 11 February) definitely cracked and the roads leading from the West were littered with Australian soldiers in all degrees of demoralization. Considerable looting of private houses, including my own, took place by these men in search of liquor. The docks were full of them and quite a large number managed to get away. The reason for this "crack" is difficult to understand as the AIF had fought extremely well in Johore and at the battle of Gemas but there is no doubt whatever that something failed. I am of the opinion that it was largely due to lack of discipline. Where discipline is weak it takes very little for a panic to set in . . . the state of discipline among the AIF was appalling. No one ever saluted, no one ever made any attempt at drill or correct turn-out. Half naked Australian soldiers roamed the streets at all hours of the day. Venereal disease, which I always consider, is a matter of discipline, had a far too high incidence among all

[89]P. Elphick, *The Pregnable Fortess,* p. 304.
[90]Ibid., p. 305.

type of troops. Lack of discipline was, I feel, a minor contributory case to the loss of Malaya. The crack in the AIF morale after the landing of February was in no small measure due to a strange absence of discipline in the AIF. I myself saw five Australian soldiers, naked except for a pair of dirty shorts, no boots or socks, lying and sprawling in the gutter of one of the main thoroughfares leading west from Singapore on 10 February. They had their rifles and were drinking from bottles. Such a spectacle reduced any confidence and respect the native population may still have retained for the white soldiers . . .

The British soldier may have lacked some discipline but he with his Indian brother at least retained their self-respect and courage.[91]

The English never really understood the Australian soldier. They appreciated his fighting potential but could never accept his attitude toward discipline. However, biased as Colonel Ashmore undoubtedly was—and we must note that his own house was looted by the Aussies, which can hardly have improved his mood—there was another, darker side to the Australian collapse:

> They were panicking, trying to commandeer boats. Dr. Albert Coates [saw] a group of Australians, displaying the independence and aggression which in other circumstances made them great soldiers, force their way at gunpoint onto one of the last boats taking out civilians.[92]

The story of the *Empire Star,* which sailed from Singapore on 11 February, supposedly carrying women and children and civilian escapees, was particularly shameful. Believing the boat to be the last to leave the harbor, Australian deserters forced their way on board, barging women and children off the gangplank and also, it was rumored but never proved, shooting a British officer who tried to stop them. The boat eventually sailed with over 150 deserters aboard, occupying space that should have been given to helpless

[91] I. Allen, *Singapore 1941–1942,* pp. 197–98.
[92] P. Elphick, *The Pregnable Fortress,* p. 306.

civilians. Some of the Australians who had failed to get aboard showed their dissatisfaction by machine-gunning the boat as it sailed.

The humiliation of the British collapse was made all the more bitter by the recriminations that followed. It seemed that everyone else was to blame. The Australian commander, General Gordon Bennett—who shares with a few unfortunate military officers the stigma of having escaped and left most of his men behind—blamed it all on the Indians. He claimed that their low morale resulted from the inability of the Eastern races to withstand the strains of modern warfare. His own argument is surely demolished by the simple thought that it was an "Eastern race" that had just defeated him. The British commander-in-chief, General Wavell was typically blunt:

> As in all other warfare, in thick or open country, in Asia or Europe, in advance or retreat, in attack or defence, the leadership of the officer and the fighting spirit of the soldier—the determination to beat the other man whatever happens—is the deciding factor. There are three principal factors in all fighting—good equipment, tactical skills and guts. But the greatest of these is *guts.*[93]

Wavell continued by adding, "For the time being we have lost a good deal of our hardness and fighting spirit." General Percival, British commander at Singapore, might have hated to hear this judgment but the facts were clear. His command had crumbled in panicky rout and Britain had suffered the worst defeat in her history because her soldiers had lacked the "guts" to hold back a Japanese army a third of their own strength. Fighting spirit needs to be renewed in each generation. If the Royal Navy had once held the world in awe, it was not sufficient for Britain's soldiers to drink their Singapore Slings and look down their noses at their Asiatic brethren. The rise of Japan

---

[93] I. Allen, *Singapore 1941–1942,* p. 200.

posed a potent threat to European domination in the Far East and what had once been won with the blood of Britain's soldiers could only be held with the same commitment of their descendants.

15

"Tackling the First Team"— The Battle of Kasserine Pass, 1943

In wartime reputations have to be earned. And reputations for tough fighting have to be paid for in blood. In the opinions of some observers—French, British, and even German—the American soldiers who arrived in Africa in 1942 did not understand this. They thought that those who talked a good war could fight one. Those who excelled on maneuvers and built up big reputations during peacetime in the United States would naturally be able to translate these achievements to the wartime environment of North Africa. The "Big Red One" was the nickname of the US 1st Infantry Division rather than the face of its commander after it had suffered heavy casualties at the hands of the Vichy forces it had come to liberate. In hushed tones new recruits were told that the Big Red One had fired the first American shots in France in 1917. It is doubtful if any of the hundreds of British, French, or German divisions on the Western Front would have included this event on their own battle honors.

The problem was that so much of America's military experience had been against the "second division." The German soldiers they had met in battle in 1918 had been war weary and the remnants of the great military machine with which Germany had begun the war in 1914. By 1918, the soldiers on the Western Front—of all nations—were either young and inexperienced—the scrapings of numerous national barrels—or old and cynical, eager to let the naive Americans do the fighting for them. After the collapse of Ludendorff's great offensive in the spring of 1918, the Germans were a spent force, fighting defensively but prone to look for ways out of the fighting in a way the Germans of 1914–16 would never have done. As a result, American "triumphs" of 1918 needed to be seen in context. And since 1918, the United States had returned to isolationism in its foreign policy, allowing its own military developments to lag well behind those in France and Germany, if not Japan and Italy.

Only Britain seemed to share the American tendency to wish to refight the last war. Even though progressive military thinkers in England, like Basil Liddell Hart and J.C. Fuller, pointed the way forward to *blitzkrieg* tactics, they were preaching to deaf ears. Only the Germans seemed to take any notice. Perhaps this was inevitable in societies essentially suspicious of the military, and eager to pass on the responsibility to more traditionally militaristic allies like France. Whatever the cause, the Americans who arrived in Africa after "Operation Torch" were arrogant, cocksure, and disdainful. The fighting men, believing that like their fathers of 1918 they were coming to redress the problems of Europe, found themselves hopelessly out of their depth in the harsh world of military professionals. If the British had taken the war too lightly at first, the terrifying efficiency of the *Wehrmacht* had soon taught them their lesson. The British had been fired in the forge of disaster—both in France and in Egypt—by superior German tactics and fighting spirit. It was a case of "adapt or die" and the British had adapted. In 1942 the Americans faced the same stark message—both in Africa against German professionalism and in the Pacific against Japanese fanaticism. It really did not impress commanders like Erwin Rommel if the "Big Red One" was coming to fight him. He would judge its men as he judged his own—by his own tough standards.

Some American officers—notably those with First World War experience—were worried about the fighting qualities of the American soldier. While the 34th Division was training in Northern Ireland during the summer of 1942, both General Eisenhower and General Mark Clark observed them on maneuvers. Clark, for one, was not happy. He felt that the soldiers "seemed fat and podgy in contrast to the lean, hard look of the British soldiers.'' Clark was concerned that the young G.I.s, drawn from the comfortable homes of the United States, would never be able to match in combat the Germans, drawn as they were from life in a totalitarian state, where discipline was absolute, where physical fitness was a virtual fetish, and where living conditions were tougher and more ''character-forming'' than in a democratic state. He decided that only the most vigorous training could bring the G.I. up to the level of the men he must fight. Unfortunately for the United States and for Clark, his colleague Eisenhower was too much of a peacetime soldier, too strong on soldiers' rights and too little on their duties. ''Ike'' even insisted that every officer should take trouble to explain to all his soldiers ''the reason for the exertion he is called upon to make . . . Any commander should be summarily relieved who neglects this important phase of training intelligent, patriotic Americans.'' Ideally, of course, Eisenhower was quite right. Perhaps the citizens of Ancient Athens would have expected to know for what they were fighting. But in an era of ''total war'' one is fortunate to be able to make a choice to take just as much of a war as may not conflict with an individual's right under the Constitution. In Russia, millions of Germans and Russians were fighting for their own survival as well as the continued existence of the states in which they lived. Large areas of Europe lay under the Nazi heel, having lost their independence because democratic leaders tried to reason with a dictator bent on world domination. It was not only American soldiers who needed to be tougher, it was their commanders as well.

The strategy of the Tunisian campaign of 1942–43 need not concern us here. Suffice it to say, that through inept leadership the American 2nd Corps under General Fredendall was guarding the Faid Pass, key to the Eastern Dorsall and gateway to western Tunisia, with inadequate strength against General Erwin Rommel's *Afrika Korps.* It was vital that the pass should be held until General Bernard Montgomery's British Eighth Army could arrive from Libya to administer the *coup de grace* to Germany's ''Desert Fox.'' But the Germans were convinced that the American troops would prove a weak point in the Anglo-American defense line and so Rommel planned to break through the Faid Pass.

It was St. Valentine's Day–14 February, 1943–but nobody except those with a macabre sense of humour was expecting a massacre. But one was about to happen. Two powerful units of the 10th Panzer Division were moving toward the Faid Pass, where General McQuillan's Combat Command A was awaiting them. The German units were headed by the massive fifty-ton Tiger tanks, against which the American Sherman and Honey tanks were virtually helpless. Colonel Waters had dug in infantry and artillery around the approaches to the pass, but these were already streaming back, concluding that they could offer no resistance to the steel monsters. Ominously, the panicky cries of "Tigers are coming!" were spreading among the American units, boding ill for the chances of better resistance farther back. Forewarned is forearmed, and even troops well back from the firing line were getting jittery. From a high point overlooking the entrance to the pass, Colonel Waters saw the Germans coming. Fifteen Honey tanks drove forward to try to delay the Germans. It was a suicidal attack. The Honeys were totally unable to face the Tigers head-on, and their 37-mm gun was thoroughly outperformed by the 88-mm cannons of the German tanks. The Honeys' attack was a mere gesture, for each of them was knocked out in turn, without scoring a single hit on their thickly armored opponents.

It took the divisional commander, McQuillan, nearly five hours to react against the German attack. Even then, what he suggested was futile. Colonel Hightower with the 2nd Battalion of the 1st Armored Regiment was ordered to go forward with his Shermans and try to stop the German advance. It was wishful thinking. Although the Shermans were far stronger than the Honeys, they had the ominous nickname of "Ronsons" owing to their startling tendency to burst into flames when hit. The American tank crews needed all the courage going to engage in battle against such odds. Some of Hightower's men fought with the courage of despair, though others looked for the first opportunity to get away. One whole reconnaissance unit of a hundred men surrendered to the Germans without firing a shot. Soon, to Hightower's chagrin, the panic spread and many other soldiers were looking for the easy way out. They simply abandoned their vehicles, took off their helmets and held up their hands. Hightower's gunners had also thrown in the towel. Abandoning their guns in the desert, they set off for the imagined safety of their own lines. The lines of the 168th Infantry were virtually swept away by fleeing men shouting, "The Krauts are coming." Officers, in the tradition of futile gestures the world over, pulled their pistols and waved them at the fleeing columns. In the words of Charles Whiting, "Here and there an officer tried to stop them, his arms outspread as if he were trying to catch his playmates at a childish game of Tag. But there was no holding them. Eyes wild and staring with fear, the fleeing men struggled on." No officer, it seemed, had the courage to fire into the deserters and demonstrate to them the consequences of running away in wartime. Because their officers would not tell them it fell to Rommel to show these young American soldiers that they were fighting for their own lives in the desert. They could never outrun a bullet.

McQuillan's thin hopes were shattered by the phone call he received from Colonel Drake, commanding a brigade of the 168th Infantry. "They're running away, General. Your men are running away." McQuillan must have known it was true but he refused to admit it. "You don't know what you're saying," he barked back. "They're only shifting position!" "Shifting position hell!" Drake replied, "I know panic when I see it." Drake was right. Just seven of Hightower's original fifty-one tanks survived the encounter.

The Americans had suffered a stunning defeat in their first encounter with the Germans. Technological superiority, notably in tank manufacture, had given the Germans a material advantage that the Americans had been unable to combat. Yet the G.I.s had run rather than confront their fears. They had shown less stomach for fighting the Germans than had the Poles at the outbreak of the war. The difference, of course, was that the Poles knew what they were fighting for—national and personal survival, in that order. In the desert the Americans could feel no commitment to the cause for which they had been sent to Europe. At that moment of crisis, democracy or the four freedoms of Franklin D. Roosevelt seemed a million miles away. The G.I.s were fighting for their own self-preservation, and if this could be achieved by running away rather than fighting, then flight rather than fight was the option every time. It was not their fault that the Germans were better at war than they were. After all, their tanks were impervious to American fire. Never mind that in Russia peasant soldiers were finding ways to halt the German juggernaut. It took fighting spirit of a different order to lay down your life in a cause other than your own. American G.I.s felt that they had too much to lose. Comfortable homes, good food, and good prospects were too much for a man to sacrifice in defense of something nobody seemed able to define. They were all conscripts anyway. After all, who would volunteer to lose his life in the Tunisian desert?

Erwin Rommel was hardly surprised at the news of his success in the battle for the Sidi Faid Pass. He knew the Americans were inexperienced and was not impressed by what he had seen of the American fighting man so far. He was hardly surprised that they were soft and prone to panic. As one German prisoner laughingly told a British Eighth Army soldier who took him prisoner, "We have the Italians; you have the Americans." Rommel believed that if he could hit the Americans hard before they were able to come to terms with modern warfare in the desert then he would be able to drive the Allies out of Tunisia. The problem for the Allies was that, after their initial setback at the Faid Pass, the Americans were already half-beaten by rumors that Tiger tanks were approaching. It would take fighting spirit of a much higher level to halt the German drive toward the Kasserine Pass, gateway to Tebessa and the Mediterranean Sea.

It fell to Colonel Stark, commander of the 26th Infantry, part of the "Big Red One," the US 1st Division, to restore America's self-esteem by taking over the Kasserine Pass and putting a stop to Rommel's progress. As General Fredendall told him, "I want you to go over to Kasserine right away and pull a Stonewall Jackson." Stark was suitably astonished: "You mean tonight, General?" "Yes, immediately," said Fredendall. Like so much about the American war effort at this time, Fredendall was all talk. Stark was not and knew that his commander was talking nonsense. Throughout Fredendall's command, more and more men were deserting, suspecting that they were no match for the Germans in spite of all the encouragement they received from their officers. Their fighting spirit was virtually nil. They were simply not ready for such desperate fighting. They felt able to take on the Vichy French; that was about their level—but not "the first team."

While Stark set off for Kasserine Pass, British Brigadier Dunphie's 26th Armored Brigade was trying to force a path through the masses of retreating Americans to reach the Pass. One of Dunphie's officers recalled passing a truck of G.I.s shrieking, "He's right behind us." Whether "he" referred to Rommel or to the Devil himself, the officer never discovered. All he learned, however, was that the G.I.s were not prepared to hold the line against Rommel until it ran up Pennsylvania Avenue in Washington, D.C. The British found their American allies distinctly frustrating. One NCO later wrote, "The road was a chaos of retreating Americans and their transport—though not enough transport to carry as many G.I.s who were all in." One of the G.I.s described his own experience:

> It was one hell of a long night. Everybody was jumpy and nervous. Some engineers came up and brought us mines to dig in. We'd never even seen a mine before and had no idea how to dig them in and place them in a pattern. But the engineers were not stopping. They were too scared. They took off and let us get on with it, digging them in the best we could in the mud. They never went off later.[94]

In the American lines chaos reigned. Patrols shot at each other and panicky G.I.s kept shouting that they could hear Tiger tanks approaching. By the morning many soldiers

[94]C. Whiting, *First Blood*, p. 158.

had quit their posts in panic. As Private Grimes commented, "If it had stayed dark any longer we'd have lost the whole goddam company!"

One thing was clear. The "Big Red One" was in trouble. First Company A of the 26th Infantry was surrounded by the Germans and its commander was captured. Then even the American Official History conceded that "the other companies went out of battalion control. Stragglers reported the situation after daylight." What actually happened was that the American forces had, almost to a man, abandoned their positions and fled at the thought of what the Germans might send against them.

While the Americans fled back from the Pass, eleven British tanks—ancient Valentines—codenamed "Goreforce" under Lieutenant Colonel Gore—struggled through the routed army to support Colonel Stark and try to hold back the German advance. What followed was another massacre. The British Valentine tanks were no match for the Tigers, but they fought until every one of them had been destroyed. They were merely a gesture against such terrifying German strength, but they were a gesture of defiance. Meanwhile, Stark's men had fought hard to hold the Germans back until Rommel's infantry had turned both American flanks and the story of the Faid Pass was repeated. Once again an American force broke completely and ran. In an atmosphere of hysteria most American soldiers took the line that "this place was too hot" and took to their heels. The defense of the Faid and Kasserine passes had been a mismatch; all part of a painful learning process for the Americans in World War II. Ill-equipped and badly trained troops of any nation would have found it impossible to withstand an elite fighting force under a gifted general with far better tanks. American fighting spirit needed to be fired in the white-hot furnace of battle, but it should be battle between equals and the G.I.s should be led by men who deserved their trust.

General Fredendall's chosen successor, General Ernest Harmon, was driving against the tide of defeat and through the fleeing American troops to reach the front. As he went he observed something he had never believed he would see. As he later wrote, "I have never forgotten that harrowing drive: it was the first—and only—time I ever saw an American army in rout. Jeeps, trucks, wheeled vehicles of every imaginable sort streamed up the road toward us, sometimes jammed two and even three abreast. It was obvious there was only one thing in the minds of the panic-stricken drivers—to get away from the front, to escape to some place where there was no shooting."

When Patton took over II Corps nobody could doubt that it was in a terrible mess. Frankly—and Eisenhower was as much to blame for this as anyone—the men had been mollycoddled. They had been brought from civilian life in the United States to a war situation and led to believe that they could conduct themselves on their own terms. They were simply "mentally and physically soft and very green." It was not their fault. Their officers had failed them. They needed to be toughened up for modern warfare; otherwise, they were merely sacrifices in a war where none but the strong survived. Patton was the man for the job. Accused of being "undemocratic and un-American," Patton regarded such accusations as compliments. He imposed a regimen that won him few friends but many admirers.

Sometimes it is easier to get the truth from one's enemies than one's friends. General Rommel had some interesting things to say about the Americans he had encountered: "The American troops could not yet be compared with the veteran troops of the Eighth Army [British], yet they made up for their lack of experience by their far better and more plentiful equipment and their tactically more flexible command. They had recovered very quickly after their first shock and had succeeded in damming up our advance." Nor were the British always as hostile toward the American soldiers as is usually suggested. Winston Churchill—admittedly half-American himself—spoke of the G.I.s as "wonderful material [who] will learn very quickly." Even after the catastrophe at Kasserine, he was not depressed: "They are brave but not seasoned troops . . . They will not hesitate to learn from defeat and will improve themselves by suffering until all their strongest martial qualities come to the front." U.S. war correspondent Ernie Pyle, who was in the thick of things with the G.I.s, reflected, "I saw them in battle and afterwards and there was nothing wrong with the American soldier . . . His fighting spirit was good. His morale was okay. The deeper he got into a fight the more of a fighting man he became." What the British regarded as indiscipline was often just high spirits, and what they thought of as weakness was sometimes a sincere idealism shared by none of the combatants in the war. Many Americans genuinely believed in the chances of a better world. Too many Europeans had looked into the pit and been scorched by the horrors they saw there. In this sense, the Americans were as desperately needed for their "clean spirit" as for their fighting spirit. Many G.I.s still spoke as if there was a cause greater than mere survival. For the British, cynicism—or a fear of their own feelings—made them reticent about why they were fighting. They assumed it was obvious: to defeat Hitler and to go home. For a lot of Americans this was only a part of it; obviously they wanted to end the war and go home, but not until the world was thoroughly clean again would they rest content. They would have been embarrassed to admit as much to their friends and comrades, but their letters home told the true story. As one private soldier explained to his folks:

> I'm in here fighting for the things freedom stands for—our church bell ringing every Sunday, the truth on our radio and in our newspapers, our children going to school and learning something beside military tactics, everyone having the same privileges to get ahead, the gang on the corner doing and saying what they please, the Stars and Stripes waving in the office yard each day, all our competitive sports without worry of doing them Hitler's way, everyone happy because he is free and doesn't have to worry because of his race and creed . . .[95]

Another soldier wrote home:

> Mom, I want you to know that I asked for combat assignment. I did so for several reasons. One is that I had certain

[95]J. Laffin, *Americans in Battle,* p. 156.

> ideals within my own mind, for which I have often argued verbally. I didn't feel right to sit safely far behind the lines, while men were risking their lives for principles which I would fight for only with my lips.[96]

Sentimental as much of this may seem, it represents a kind of courage, firm but generous, that is genuinely American. It was the sort of spirit that could carry men through early setbacks like that at the Kasserine Pass and on to the greater challenge of D day and the Normandy campaigns and finally to triumph and quiet resolution in a defeated Germany.

[96]J. Laffin, *Americans in Battle,* p. 156.

16

"Tell It to the Marines"— The Invasion of Tarawa, 1943

The American assault on the atoll of Tarawa in the Gilbert Islands in 1943 resulted in one of the bloodiest single days in the history of the United States Marines. In fact, in the words of the official historian, "Tarawa was the most heavily defended atoll that would ever be invaded by Allied forces in the Pacific. With the possible exception of Iwo Jima, its beaches were better protected against a landing force than any encountered in any theater of war throughout World War Two."

The battle fought on the island of Betio, part of the Tarawa Atoll, from 20 to 23 November, 1943, was remarkable not only for the courage shown by the soldiers on both sides but by the clash of cultures that took place. Within the culture of the United States, the men who went ashore at Tarawa displayed bravery in the best traditions of the marines, a service that could trace its history back through all the wars the Republic had ever fought. The American marines fought as Americans, as marines, yet ultimately as individuals, prey to the hopes and fears of men brought up within an unmilitary Anglo-Saxon, Christian tradition of democracy and freedom of the individual. The Americans on Tarawa had the freedom to think as individuals, to make choices, to fight or flee, to accept the responsibilities of their actions and, ultimately, to live or die by them. They fought for their country, for the marines, and for their friends, yet they fought without fanaticism. Above all, they fought to live and survive rather than to die.

The Japanese soldier, on the other hand, possessed neither the freedoms felt by his American opponent nor the strong individual responsibilities. In a sense, he had already given his life to his emperor, whom he regarded as a god. For him survival in battle was merely proof of his success as a soldier, it was not an end to be sought if it conflicted with his duty. He saw honor in death in battle, perhaps even a way of ensuring the afterlife. The result was that surrender to an enemy was a violation of everything he held most important. He could confidently fight to the death in the knowledge that it was required of him by both church and state—by the emperor as god. This made the Japanese soldier a doughty opponent in battle. At first, Anglo-American soldiers—including Australians and New Zealanders—found this a frightening aspect of the Japanese. They appeared to hold their lives so cheaply that there could never be an easy victory against them. Whereas, in European campaigns, at the final gasp, soldiers of every nation would seek to surrender to save themselves from unnecessary slaughter, the Japanese were guaranteed to fight to the last man. This fanaticism ensured heavy casualties and might deter enemies from combat if it could be avoided. There was nothing new in such fanaticism. During *Jihad,* Muslim warriors would fight without fear of death, knowing that if they should fall in battle, they would achieve paradise. Centuries of crusading warfare convinced Christians of the severity of religious wars. Vikings were said not to fear death as a warrior killed in battle would go to Valhalla. Yet such fanaticism was distinctly and frighteningly different for twentieth century Americans, most of whom had little or no experience of war and who had been brought up to value life, their own and that of others. Where the Japanese soldier would commit suicide rather than lose honor by surrendering, or even kill those of his fellows who were in danger of being taken alive, the American marine would fight to preserve his life in combat. For different reasons this made both formidable fighters. And if one removes the possibility of retreat for the American or survival for the Japanese, one is creating a scenario in which both will fight with the utmost ferocity.

The Japanese defenses on Tarawa Atoll, notably on the main island of Betio, which was where the airfield was situated, were quite astonishing for an island measuring no more than 4,000 yards by 500, and which was nowhere more than ten feet above sea level. It had been turned into a fortress island with a garrison of 2,619 men along with engineers and technical workers who would join the battle if needed. Palm logs linked to sand emplacements ran along the seafront, along with covered trenches and massive concrete bunkers. In addition, the island bristled with all kinds of guns, from 8-inch naval guns removed by the Japanese from the harbor at Singapore, to masses of heavy machine guns. There were also fourteen light tanks, no match it seemed for American Shermans, yet in the event surprisingly successful in the fighting ahead. The Japanese commander, Rear Admiral Keichi Shibasaki, was so confident of his ability to repel any attack from the sea that he boasted that a million men could not take the island in a thousand years. Ignoring the poetic license, the Americans planned not so much to "take" the island as to "steamroller that place until hell wouldn't have it," in the words of one officer. Admiral Howard Kingsman, in charge of naval gunfire support, was as confident as Shibasaki. He told the marines before their assault, "We will not neutralize; we will not destroy; we will obliterate the defenses on Betio!" It was all very well for middle-aged senior officers to rattle their sabers and make threats, but it was young men who were going to have to test out the truth of their boasts. And the truth was that ships had rarely proved effective against land fortifications in the past owing to the angle of fire of naval guns, which were designed for ship-to-ship encounters, usually conducted at long range. Even though the Americans were going to use the combined fire of three battleships, six cruisers and nine destroyers, the results were going to be very disappointing. Although the presence of the "battlewagons" may have inspired the waiting marines with confidence it was a false one as they were soon to discover.

The assault troops would consist of the 2nd Marine Division, under Major General Julian Smith, which consisted of 20,000 men, made up of three infantry regiments (2nd, 6th, and 8th), each of three 1,000-strong rifle battalions and a heavy weapons company in support. The first problem they faced—and a vital one as it turned out—was the state of the tide at Tarawa, which was very difficult to predict. The navy was convinced that there would be enough water covering the reef for shallow-draft landing craft to easily pass over it. However, a long-standing British resident, Major Frank Holland, did not agree with them. He warned the Americans there would not be enough water over the reef until December, and if they tried to use landing craft, they would rip their bottoms on the reef or get stuck. Whom to believe? The US Navy or Major Holland? Unfortunately, most people got it wrong: they went along with the navy. Anyway, nobody had time to wait until December to see if Major Holland was right. Luckily, some of the marines were privately listening to Holland. After all, it was their feet that would get wet. As a result, Colonel Shoup insisted on a new type of assault boat that could pass over the reef in shallow water. These were still at the developmental stage, but Shoup was insistent; he would give them a go. Then came the bad news; none was available. Shoup now decided to use amtracs (a tracked amphibian vehicle designed for use in the Florida Everglades) instead. A compromise was reached with the navy. The first three waves in would use amtracs; later waves would use the naval landing craft.

Naval opinion was that the marines were fussing over nothing. After they had finished bombarding the island the Japanese would be in no condition to resist an assault. The bombardment duly took place. It was impressive—indeed awesome to the marines watching—but it was ineffective as events were going to show. The whole of Betio was covered in a cloud of smoke. Now it was the turn of the amtracs to make their three-mile approach to the beaches. They passed over the reef, scraping their keels (Holland had been right, then), with the naval guns still pouring high explosive onto the Japanese positions. Betio seemed to be nothing but an inferno of explosions and raging fires. How could anyone live through such a storm of fire? The Japanese must be dead or disabled, the marines reasoned. Perhaps the navy had got it right after all. Showing incredible courage Lieutenant William Hawkins and his company landed on the pier, eliminating the Japanese gunners who might have opened a flanking fire on the amtracs. Unfortunately, Hawkins's raid on the pier was about the only part of the plan that worked, for, with a startling suddenness, the amtracs came under a storm of fire from Shibasaki's machine gunners and antiship gunners. Private Baird described his landing:

> Bullets pinged off that tractor like hailstones off a tin roof. Two shells hit the water twenty yards off the port side and sent up regular geysers. I swept the beach [with a machine gun] just to keep the bastards down as much as possible. Can't figure out how I didn't get it in the head or something.
>
> We were 100 yards in now and the enemy fire was awful damn intense and gettin' worse. They were knockin' boats out right and left. A tractor'd get hit, stop, and burst into flames, with men jumping out like torches . . . Bullets ricocheted off the coral and up under the tractor. It must've been one of those bullets that got the driver. The boat lurched and I looked in the cab and saw him slumped over. The lieutenant jumped in and pulled the driver out, and drove himself until he got hit.
>
> That happened about thirty yards offshore. A shell struck the boat. The concussion felt like a big fist—Joe Louis maybe—had smacked me right in the face. Seemed to make my face swell up. Knocked me down and sort of stunned me for a moment . . . My assistant, a private with a Mexican name, was feeding my gun, had his pack and helmet blown right off. He was crumpled up beside me, with his head forward, and in the back of it was a hole I could put my fist in. I started to shake him and he fell right on over.
>
> Our boat was stopped and they were laying lead to us like holy hell. Everybody seemed to be stunned, so I yelled, "Let's get the hell out of here."
>
> I grabbed my carbine and an ammunition box and stepped over a couple of fellas lying there and put my head up. The bullets were pouring at us like a sheet of rain . . . Only

> about a dozen out of the twenty-five went over the side with me and only about four of us ever got evacuated.[97]

Many of the marines in the first wave were killed in their amtracs, while others who leaped overboard drowned in the chaos. Eight of the amtracs were sunk, while those that survived were riddled with heavy machine gun fire. Some amtracs just wallowed about in the waves, full of dead marines. Some that reached the beach were just flaming wrecks. But worse was to follow. The fourth wave of Marines were traveling in naval landing craft and they ran aground on the reef. There was no option for the marines but to jump over the side and wade to the beach seven hundred yards away, all the time an easy target for the Japanese gunners. The water was chest-high to begin with and this offered some protection, but as the ocean floor sloped upward they found themselves exposed to enemy fire. The Japanese were exultant. They could hardly have envisaged such a target. The American invaders were helpless. Near the reef was an abandoned steamer, which the Americans had overlooked in their planning. Unfortunately, it was filled with Japanese machine gunners, who fired into the backs of the marines as they struggled ashore. Wounded men drowned, out of reach of help. Soon the waves were burdened with dozens of bodies, tossed to and fro onto the white sand. The terrible journey to the shore had gone close to shattering the morale of the assault troops. Everything had gone wrong and now they were trapped on the beach, pinned down by heavy enemy fire. But there was no turning back; death was behind as well as ahead. Young marines showed incredible courage and determination. One young soldier on Beach Red 3 came in with half his face slashed to ribbons. The doctor slapped a large bandage on his wound and tried to get him to lie down, but the adrenaline was pumping and he would not rest. "Ah, fuck that!" he said, spitting blood, "I gotta get back to my outfit."

[97]C. Gregg, *Tarawa,* pp. 70–71.

Once the assault waves had gone in, it was decided to land the Sherman tanks. But the miscalculation about the reef meant that the Shermans had to drive ashore themselves from the reef, led by guides on foot, all the time exposed to enemy fire. Many guides fell but, to their eternal credit, volunteers thrust themselves forward for this near-suicidal duty. Never had the fighting spirit of American servicemen been seen to greater credit. Yet for many of the marines, fear stalked the island as Private Grogan described:

> It was like being completely suspended, like being under a strong anaesthetic, not asleep, not even in a nightmare, but just having everything stop, except pain, and fear, and death. I was afraid. Everybody was afraid. And no one was too proud to admit it. Our lips cracked with the dryness of fear, and our voices sounded to us like voices of complete strangers, voices we had never heard before. By the second day men's mouths were literally black with dryness from fear. Not just a few of them, but all of them. Their faces were black with soot. Their eyes were inflamed from the smoke, from the sun and, also, from fear.[98]

The battle had become a struggle between riflemen on both sides. Small groups rushed forward from the beach to find cover, attacking the enemy strongpoints with grenades and flamethrowers. The Japanese troops in forward positions were already committing suicide to avoid capture.

On the second day General Julian Smith came ashore to supervise the fighting and replace Colonel Shoup, who was out on his feet after fifty hours of combat. Yard by yard the Japanese positions were taken until by the end of the second day the Japanese were confined to just one end of the island. Their final moments had come. During the night of 22/23 November, most of the surviving Japanese organized a suicide attack which reached the marine lines, but ended with almost every defender dying at the hands of the American soldiers. The Japanese attacked screaming

[98] W. Richardson, *The Epic of Tarawa,* p. 87.

"Banzai," or "Marine, you die," or "Japanese drink marine's blood." It was to no avail. Their battle cries had a hollow ring. There were just 146 survivors from the Japanese garrison, most of them Korean workers, and many of the Japanese had committed suicide in their pillboxes. Casualties on the American side had been heavy: 51 officers and 853 men had been killed in the fighting, with 109 officers and 2,124 men injured. Of the 2nd and 8th Marines more than 35 percent had become casualties. Of the amtracs, 35 had been sunk and 26 disabled by gunfire. It had been a flawed action on the part of the navy planners. But the marines had made up for every planning deficiency and overcome every obstacle. Their fighting spirit had been inspiring and they had turned a potentially disastrous operation into a great victory.

17

Into the Dragon's Lair—The Battle of Hürtgen Forest, 1944

In terms of its pitiless terrain and its morale-sapping effect on troops, this battle has become known as the "American Passchendaele." Conceived in folly and pursued with a relentless determination on the part of American commanders the "green hell of Hürtgen" tested the fighting spirit of the American G.I. to its limit and, in many cases, far beyond it. In the words of General James Gavin, "For us the Hürtgen was one of the most costly, most unproductive and most ill-advised battles that our army has ever fought."

There was something Wagnerian about the Hürtgen Forest. In fact, the Germans had fortified it so strongly that in their own words it resembled a "dragon's lair." It positively bristled with mines, imaginatively dubbed "dragon's teeth" by the defenders, who occupied massive concrete pillboxes, impervious to anything but a direct hit from the air. To attack such defenses head-on was tantamount to replaying the Charge of the Light Brigade. But American commander "Lightning Joe" Collins was determined to pull the dragon's teeth out by the roots and it did not seem to matter how many American lives were lost in the process. Although the target was of no great strategic value, the prestige of the American army was at stake. It was at a critical stage in the war. Most informed observers had expected the German military machine to collapse once the Anglo-American forces had succeeded in breaking out of their beach heads in Normandy. Instead, the Germans conducted a skillful and dogged defensive campaign all across northern France, sapping the morale of the Allied troops and demonstrating just how hard it was going to be to stage an immediate invasion of Germany itself.

What sort of men were being sent to fight in the impossible terrain of the Hürtgen Forest? On the German side, the joke ran among the G.I.s, the barrel had been well and truly scraped, so that the forest was garrisoned by decrepit veterans of the First World War, "stomach cases," cripples with glass eyes, and men with wooden legs. But the Americans who had fought their way to the German frontier from the shores of Normandy were not amused. As one exclaimed in frustration, "I don't care if the guy behind that gun is a syphilitic prick who's a hundred years old—he's still sitting behind eight feet of concrete and he's still got enough fingers to press triggers and shoot bullets." And that was just the point. All the advantages lay with the defenders. As a German general later observed, "The fighting in the wooden area denied the American troops the advantages offered them by their air and armored forces." For some reason, known only to the High Command, G.I.s were being sent into the forest on some kind of macho exercise to prove that man for man they were a match for the Germans. But given the conditions they were never going to be a match—even for the scrapings of the German barrel.

Soon the confidence that had sustained the G.I.s in their race across northern France began to decline. The fighting in the forest affected morale and more and more psychiatric casualties occurred. The gloomy atmosphere in the forest and the sense of isolation played havoc with men's minds. The historian of the US 9th Division observed, "With adverse weather conditions and the impossibility of continued and accurate artillery or air support, many soldiers felt as if they were fighting in the dark. Each infantryman, moreover, was on his own. More than at any other time GIs and officers experienced the tension and strained nerves that make men victims of combat fatigue." And the stress was no respecter of rank. Even battalion commanders were subject to depression so deep that it manifested itself in weeping fits and refusal to fight. At times it seemed as if the 9th Division was suffering from a collective nervous breakdown. Under intense and accurate bombardment soldiers simply cracked. One later remembered the effect:

> Me and this buddy of mine were in the same hole with only a little brush on top and I remember I was actually bawling. We were both praying to the Lord over and over again to please stop the barrage. We were both shaking and shivering and crying and praying all at the same time. It was our first barrage.
>
> When it stopped both of us waited for a while and then we crept out of the hole and I never saw anything like it. All the trees were torn down and the hill was just full of holes . . .
>
> They sent me back to an aid station for a while and I guess they treated me for shock or something. Then they sent me back to my unit . . . And it was the same shells, the same goddamned shells. Soon as I got there, the Jerries started laying them on again. They started laying them all over the road and I tried to dig in and then I started shaking and crying again . . . I guess I must have gone off my nut.[99]

It was shell shock and the G.I.'s account would have been echoed by millions of soldiers in both world wars.

By the time the 9th Division was withdrawn from the Hürtgen Forest casualties were running at more than thirty percent and morale was a developing problem. Combat fatigue was widespread and a rash of self-inflicted injuries were beginning to appear as symptoms of a far deeper malaise. It was not just the enemy that was sapping the men's fighting spirit, it was the forest itself. Winter was setting in and terrain that had previously been difficult became almost impossible. The fighting in the forest assumed a primitive character, a hand-to-hand infantry slogging match, where technology held few advantages.

To replace the 9th the choice was the 28th Infantry Division, a unit with a long and proud record going back to the War of Independence. Unfortunately, many of its se-

---

[99]C. Whiting, *The Battle of Hürtgen Forest,* p. 25.

nior officers were as much a part of its long history as its tattered battle flags. They were too old and lacked the energy necessary to lead a division into the sort of bitter fighting that was taking placed in the Hürtgen Forest. After a general clear-out of "deadwood," command of the 28th was eventually given to Norman "Dutch" Cota, a fighting general if there ever was one, who was reputed to have saved the day during the carnage on Omaha Beach and was later wounded near St. Lô. Interestingly enough, at a time when senior officers were encountering more and more cases of shell shock and were acknowledging the views of psychiatrists on the nervous complaints of frontline soldiers, "Dutch" Cota was no believer in combat fatigue. He believed in rest as a cure-all. For Cota, every human right carried with it a consequent obligation. As he told his men, "For every right that you enjoy there is a duty that you must assume. You've heard a lot of talk about rights. Now you'll hear a lot about duty."

Perhaps Cota suited the situation in which the Americans found themselves in the fall of 1944. The ordinary G.I. taken straight out of civilian life and trained to be a killer, was probably too soft. Both nature and nurture worked against the army drill sergeants. Compared with German or Japanese soldiers, whose traditional military virtues of obedience and sacrifice were anathema to Anglo-Saxon values of freedom and individualism, or with the tough and often brutal Russian peasant soldiers, the American soldier was a product of a society where human rights were enshrined in the Constitution. Americans had traditionally mistrusted standing armies.

The British had already encountered the same problem in 1941. Sir John Kennedy had observed after the fall of Singapore:

> We had cause on many previous occasions to be uneasy about the fighting qualities of our men. They had not fought as toughly as the Germans or the Russians, and now they were being outclassed by the Japanese. There were two reasons for this. The first was that we had only begun to form our army in earnest after the war had broken out,

> and it may be accepted that it takes about three years to organize and train troops, and to produce modern equipment for them. The second reason was that we were undoubtedly softer, as a nation, than any of our enemies, except the Italians. This may be accounted for by the fact that modern civilization on the democratic model does not produce a hardy race, and our civilization in Great Britain was a little further removed from the stage of barbarity than were the civilizations of Germany, Russia and Japan.[100]

As a result, there were serious disciplinary problems in the American army in France, and it has been estimated that there were 20,000 American deserters in Paris alone. Perhaps this was inevitable in view of the sheer size of the organization, but it was more likely a result of the expectations that soldiers had of army life and the obligations that the authorities felt for every mother's son. German and Japanese soldiers—let alone the Russians—would not have believed the privileges that the G.I.s enjoyed during R & R (rest and recreation). Sometimes R & R stood for little more than another chance to contract VD—incredibly, a battalion's worth of soldiers was missing daily for attendance at the "pox hospital"—and, as a result, the combat efficiency of the US Army in and around the Hürtgen Forest was seriously undermined.

On 2 November, the 28th division began its assault on the village of Schmidt, in the middle of the Hürtgen Forest. Fourteen thousand infantrymen were involved, but most of them were replacements and a lot were "green" troops. At once everything went wrong that could go wrong. The one thing green troops needed was belief; belief in their commanders and belief in the efficiency of their own artillery and air support. Their fighting spirit was unproven. It was still an ideal rather than a reality. It existed in principle and would do so in practice provided that everything went according to what they had been taught. What green troops do not need is an outbreak of amicide. But this is exactly what they got. Shaken by a friendly air attack, the G.I.s

[100]Sir J. Kennedy, *The Business of War,* p. 198.

stumbled into well-concealed German machine gun fire. Casualties mounted rapidly and the morale of the young soldiers collapsed. Discipline began to waver and men started leaving the firing line. At first just a few left and then, like a widening stream, dozens of G.I.s, unable to take it any longer, set off in a rout. Combat fatigue was spreading in the 28th like bubonic plague.

The sight of the German tanks, as impervious to their own mines as they were to the feeble American bazookas, crashing through the undergrowth toward them was just too much for the inexperienced men of the 3rd Battalion of the 112th Regiment. They began to fall back without orders and then broke in panic. Officers who tried to stem the rout were swept away. The news was quickly relayed to General Cota. He could hardly believe what he was hearing. How could a victory as complete as his be overturned in a matter of hours. Surely it must be a mistake, just a temporary, local difficulty. But this was neither temporary nor local. In the words of Charles Whiting, historian of this ill-advised campaign:

> Everything now started to go radically wrong for Cota's 28th Infantry Division. The sense of panic and hopelessness was infectious, as the wide-eyed, ashen-faced stragglers from the shattered 3rd started to come through the next village, Kommerscheidt, held by the 112th's First Battalion. The defenders began to grow nervous and apprehensive. Officers and noncoms tried to hold the fugitives, kicking and striking them. More than once an officer drew his .45 and threatened them, but as S. Sgt. Frank Ripperdam reported later, "There was no holding them. They were pretty frantic and panicky." In the end only some two hundred were coerced into staying in the line at Kommerscheidt; the rest fled into the woods.[101]

Even as the fighting spirit of many of the American infantry was disintegrating, individuals were reacting differently. Grouped around three Sherman tanks in Kommerscheidt,

[101]C. Whiting, *The Battle of Hürtgen Forest,* p. 80.

some elements of the 112th hung on. But hanging on was not enough for Norman Cota. He wanted an immediate counterattack to regain Schmidt. He had promised instant success and now found himself in the embarrassing situation of expecting a visit the next day from "Ike" himself, just at the moment that his men were cracking and running. But the counterattack existed only in Cota's mind. The reality was that on the ground his infantrymen were shattered. As Whiting wrote, "Some were crying as if broken-hearted. Others just slumped apathetically and had to be ordered to eat their rations." Nor was their battalion commander, Colonel Hatzfield, any stronger. He had looked into the abyss and had lost all hope. He was a broken man.

Far from the 28th Division counterattacking, it was the Germans who seized the initiative and the Americans who cracked. First individuals threw down their weapons and ran, then whole companies. Panic soon set in and even company commanders joined the headlong flight to the rear. Lieutenant Condon of E Company later commented:

> It was the saddest sight I have ever seen. Down the road from the east came men from F, G and E Companies: pushing, shoving, throwing away equipment, trying to outrace the artillery and each other, all in a frantic effort to escape. They were all scared and excited. Some were terror-stricken. Some were helping the slightly wounded to run, and many of the badly wounded, probably hit by artillery, were lying in the road where they fell, screaming for help. It was a heartbreaking, demoralizing scene.[102]

The staff officers who tried to stop the panicky retreat reasoned with the men that the Germans were not following up their initial breakthrough but nobody wanted to listen. Meanwhile, Colonel Hatzfield had collapsed, a victim of complete combat exhaustion, and had been replaced. For the second time in two days a famous American regiment had cracked so completely that even its officers had fled with their men.

---

[102]C. Whiting, *The Battle of Hürtgen Forest,* p. 83.

It was clear to General Cota that he could not regain the village of Schmidt with the resources available to him. The fighting spirit of the 28th was too low. The 112th Regiment had cracked. Yet the prestige of the American army was at stake. And prestige was closely entwined with the fighting spirit of the individual soldier. G.I.s would fight well because they believed they were part of something fine. Take away that belief for a moment and so much would be lost that every future battle would become that much more difficult. The enemy would take comfort in the knowledge that in the Hürtgen Forest the Americans had cracked and run, officers leading their men to the rear and joining the pell-mell rush for safety. *Esprit de corps* might be difficult to define, but every soldier needed to feel that he belonged to a unit of which he could be proud. This was essential to his morale and ultimately to his value as a soldier.

Cota should have stood out for a complete withdrawal of his battered division. It would have been an admission of defeat, of course, and it would have taken the sort of moral courage that was in short supply among the American top brass at this difficult time. Instead he allowed himself to be talked into a renewed assault by the exhausted but as yet unbroken 110th Regiment. So for five more days the martyrdom of Norman Cota's 28th Division continued in the Hürtgen Forest. By the time the 28th was eventually withdrawn it had suffered more casualties than any American division in a similar action in the entire war. The shattered 112th had suffered more than sixty-six percent casualties.

It is hardly surprising that the only case of a G.I. shot for desertion or cowardice occurred at this time. The case of Eddie Slovik is dealt with elsewhere (see Heroes and Villains). Yet Slovik's sacrifice was an irrelevance in view of the massive collapse of morale on both sides during the Hürtgen fighting. The Germans might have been defending their own hearths and homes, but this did not prevent a collapse of fighting spirit, which had to be remedied in ways far more drastic than the execution of a single infantryman. Desperate situations demanded desperate remedies. And with the Soviet forces in the east driving through Poland and threatening the eastern borders of the Reich, with Field Marshall Alexander's Anglo-American forces emerging from the Italian peninsula and threatening the "soft underbelly" of Germany, and Generals Montgomery, Bradley, and Patton forcing their way into Germany from the west, situations came no more desperate than that facing the Third Reich at the end of 1944. There could be no weakness from her soldiers, certainly no combat fatigue. Men who ran away were cowards and were unworthy of their heritage. Ten thousand such deserters were shot without even a trial. Many more were given the chance to redeem themselves by returning to duty in a punishment battalion—actually a suicide squadron—who were sent on missions of such danger that few ever returned. Two such battalions—the 333rd and the 999th—were specially designated to undertake "Ascension Day Missions." Men who had broken down in battle were allowed to die in action rather than against a wall facing a firing squad. It all sounds like the worst excesses of a totalitarian regime, yet Eddie Slovik might have preferred a "one-way ticket" to the ignominy of his judicial murder.

The battle ground on, mincing up the manpower of both sides. The Germans were reduced to bringing in the boys of the *Volksgrenadier* divisions. These lads for all their youth and inexperience had the fanaticism of a Nazi upbringing in the Hitler Youth and fought with a fury more reminiscent of the Japanese rather than the mature German soldier. It was a doubly difficult experience for the G.I.s: having to kill boys who would kill you if you indulged in a moment's sentimental thought about them. As Major Freeman wrote, "the Generals had no idea what the men were up against . . . Someone seemed to think it was a Fort Benning exercise instead of a penetration of a thickly mined, well fortified dense forest where you were lucky if you could see twenty feet." But Freeman was wrong. It was not so much that the generals did not know what the G.I.s were up against; it was more a case that they did not care. They could not afford to care. War was hell; everybody knew that. And in war there would be casualties; thousands of them. They could not be expected to fight a modern war against a military giant like Germany if they had to worry if every mother's son was being tucked in every night. It sounded callous; it was callous; but it was war. In any case, the United States was a big country and had a big population. There were always more men to take the place of those who didn't make it. Yet the generals were wrong as well. Trained infantrymen were in short supply. Certainly, America could go on sending cannon fodder like Eddie Slovik to France to die for their country. But it was just murder. Men like Eddie Slovik were not equipped mentally or physically to fight at the front. They would either break down or they would run. And how would the sight of more and more unsuitable soldiers running in terror help the morale of the rest?

The problem of manpower at the front was a pressing one for the American High Command. In the last days of 1944 it was estimated that the army was losing up to 3,000 men a day—through death, injury, breakdown or desertion—and was in return receiving just one thousand replacements. Even this was being achieved by starving the Pacific theater of troops and by sending out men unsuitable for combat. Moreover, discipline was breaking down in many units, with self-inflicted injuries becoming widespread. In the time-honored tradition of armies in both world wars, men who wanted out shot off their thumbs or their big toes or wounded their left hands. They generally asked their friends to shoot them to confuse investigations, or fired through a loaf of bread to prevent powder burn from close-range firing. It was all a desperate game. If they were caught, the most they could suffer was six months in prison for carelessness. Many G.I.s thought the forfeit well worth paying. Anything was better than the horror of fighting on in the Hürtgen Forest. Yet most shirkers preferred desertion to self-mutilation. Thousands simply walked away from their units and found refuge in the big cities nearby, like Paris or Brussels. Soon each city boasted its own American quarter, made up of American deserters. The less ambitious lived nearer their units in caves and hovels made from logs, subsisting on the canned foods they could steal from the camps.

The problem of low morale, unsuitable replacements, and deserters was in itself a self-inflicted wound from a pistol fired by the US Army itself. Its system of distributing manpower was archaic and in many ways counterproductive. And one thing was certain: it did nothing to boost the fighting spirit of the G.I.s already in Europe. Essentially, the system of allocating recruits to the infantry was fatally flawed. It may have been a simplistic assumption on the part of the recruitment services in other countries, but it was generally assumed that the infantry actually *fought* the enemy. They did not repair planes, cook meals for officers, file inventories, drive senior offices from one cocktail party to another or report G.I.s for being late returning from a furlough. The infantry carried their rifles into action, fitted a bayonet to them, and often met the enemy in hand-to-hand battle. It was rough and frightening, and it required men of physical strength and courage. It needed men who could give and take a blow, who could kill a man by plunging a bayonet into his belly and twisting it. The British insisted that an infantryman passed for active service had to belong to the top category of physical fitness; the Americans asked only that the infantryman was fit for "general" service. As a result many G.I.s were taken into the infantry who were basically afraid of the idea of close combat with an enemy, but could quite easily have handled any assignment that kept them out of the firing line. It was absurd that the military police was filled with muscular men who were quite happy to take on any amount of drunk and disorderly soldiers on a Friday night, while the front line contained too many men who were happier lecturing a class of teenagers on French verbs or keeping the books in an actuarial office. The overriding question on personnel selection for the front line was: could the men selected fight, march, and survive in battle conditions? And the answer in too many cases was a negative one. The resultant effect on morale was devastating. Placing square pegs in round holes was one thing, placing soft civilians in situations requiring hard-bitten fighters was quite another. In army circles it was believed that men who were physically fit must be suitable to fight for their country. In principle this was all very well. In practice, however, too many American civilians were soft compared to the Germans and Japanese they faced. By 1943, replacements tested at Fort Meade were found to be shorter, lighter, and weaker than the average for the existing army. It was the same problem that the British had found in 1914, notably with working-class men from industrial backgrounds. And there were no short-term solutions. The only answer was for the authorities to ensure that the right men were in the right place at the right time. No more Eddie Sloviks should have been sent to the front line to fail. Yet the system stayed the same, and teenage G.I.s were sent into action after just thirteen weeks training. Later research showed that up to fifty percent of all replacements sent to the front subsequently admitted that even in action they had not fired their rifles, either through fear or through incompetence.

While Eisenhower and his lieutenants brooded over the failure of American youth and the breakdown in morale right across the front, the Germans were planning a surprise lightning strike in the Ardennes, which was to be known to history as the "battle of the Bulge." Yet after the horrors of the Hürtgen Forest the American army was to rediscover itself in this new and terrible conflict against an enemy assumed to have been defeated. While the Germans fought with a spirit born of desperation, many Americans recaptured their pride and their patriotism in this epic *Götterdämmerung.*

They are surely to be esteemed the bravest spirits who, having the clearest sense of both the pains and pleasures of life, do not on that account shrink from danger.

—THUCYDIDES, *History of the Peloponnesian Wars*

Bibliography

The following works have been most useful to me during the writing of this book and I have referred to many of them in the text. I acknowledge the great assistance I have gained from those books marked with an asterisk.

L. Allen, *Singapore 1941–42,* London, Davis Poynter, 1977.

A. Babington, *For the Sake of Example,* London, Leo Cooper, 1983.

N. Barber, *Sinister Twilight,* London, Collins, 1969.

*M. Barthorp, *Heroes of the Crimea,* London, Blandford, 1991.

J. Baynes, *Morale: a Study of Men and Courage,* London, Cassell, 1967.

C.E.W. Bean, *Official History of Australia in the War of 1914–18,* Sydney, Angus and Robertson, 1942.

E. Bergerud, *Red Thunder, Tropic Lightning,* Boulder, Westview, 1993.

O. Bradley, *A General's Life,* New York, Simon & Schuster, 1983.

M. Brown, *Tommy Goes to War,* London, Dent, 1978.

C.C. Buel and R. U. Johnson, *Battles and Leaders of the Civil War* Volume 3, New York, Century, 1888.

C. Carlton, *Going to the Wars,* London, Routledge, 1992.

G. Chapman, *Vain Glory,* London, Cassell, 1968.

P. Charlton, *Pozières 1916,* London, Leo Cooper, 1986.

H.S. Commager, *The Blue and the Gray,* 2 Volumes, Indianapolis, Bobbs-Merrill, 1950.

*J. Costello, *Love, Sex and War,* London, Collins, 1985.

F.P. Crozier, *A Brass Hat in No Man's Land,* London, Michael Joseph, 1930.

F.P. Crozier, *The Men I Killed,* London, Michael Joseph, 1937.

Carlo D'Este, *Decision in Normandy,* London, Collins, 1984.

Blaise de Monluc, *Military Memoirs,* (ed. I. Roy), London, Longman, 1971.

*E. Dinter, *Hero or Coward,* London, Frank Cass, 1985.

C. Duffy, *The Military Experience in the Age of Reason,* London, Routledge, 1987.

R.L. Eichelberger, *Our Jungle Road to Tokyo,* New York, Viking, 1950.

*J. Ellis, *The Sharp End of War,* London, David and Charles, 1980.

P. Elphick, *The Pregnable Fortress,* London, Hodder and Stoughton, 1995.

C. Falls, *Caporetto 1917,* London, Weidenfeld and Nicholson, 1966.

P. Fussell, *Wartime, Understanding and Behaviour in the Second World War,* Oxford, Oxford University Press, 1989.

M. Glover, *The Peninsular War,* London, David and Charles 1974.

*M. Glover, *That Astonishing Infantry,* London, Leo Cooper, 1989.

R. Graves, *Goodbye to All That,* London, Penguin, 1960.

C. Gregg, *Tarawa,* New York, Stein and Day, 1985.

P. Griffith, *Rally Once Again,* London, Crowood, 1987.

V.D. Hanson, *The Western Way of War,* Oxford, Oxford University Press, 1990.

M. Hastings, *Overlord,* London, Michael Joseph, 1984.

*P. Haythornthwaite, *The Armies of Wellington,* London, Arms and Armour, 1994.

C. Hibbert (ed.), *The Recollections of Rifleman Harris,* London, Leo Cooper, 1970.

V. Hicken, *The American Fighting Man,* New York, Macmillan, 1969.

*R. Holmes, *Firing Line,* London, Penguin, 1987.

*A. Horne, *To Lose a Battle,* London, Penguin, 1979.

*W.B. Huie, *The Execution of Private Slovik,* London, Jarrolds, 1954.

E. Junger, *Storm of Steel,* London, Chatto and Windus, 1929.

*J. Keegan, *The Face of Battle,* London, Penguin, 1978.

*J. Keegan and R. Holmes, *Soldiers,* London, Hamish Hamilton, 1985.

J. Laffin, *Jackboot,* London, Cassell, 1965.

J. Laffin, *Americans in Battle,* London, Dent, 1973.

L. MacDonald, *1914,* London, Penguin, 1987.

L. MacDonald, *1915,* London, Headline, 1993.

L. MacDonald, *Somme,* London, Michael Joseph, 1983.

L. MacDonald, *They Called it Passchendaele,* London, Michael Joseph, 1978.

T.H. McGuffie, *Rank and File,* London, Hutchinson, 1964.

H. McManners, *The Scars of War,* London, Harper Collins, 1993.

C. Mercer, *Journal of the Waterloo Campaign,* London, n.p., 1877.

*C. Messenger, *For Love of Regiment,* London, Leo Cooper, 1994.

*M. Middlebrook, *The Kaiser's Battle,* London, Penguin, 1983.

*M. Middlebrook, *The First Day of the Somme,* London, Penguin, 1984.

*W. Moore, *See How They Ran,* London, Leo Cooper, 1970.

W. Moore, *The Thin Yellow Line,* London, Leo Cooper, 1974.

*Lord Moran, *The Anatomy of Courage,* London, Constable, 1945.

T. Norman, *The Hell They Called High Wood,* London, Patrick Stephens, 1984.

*D. Omissi, *The Sepoy and the Raj,* London, Macmillan, 1994.

G.S. Patton, *War As I Knew It,* London, W.H. Allen, 1948.

B. Perrett, *Last Stand,* London, Arms and Armour Press, 1992.

*B. Pitt, *1918: The Last Act,* London, Cassell, 1962.

D. Prior and T. Wilson, *Command on the Western Front,* Oxford, Blackwell, 1992.

*J. Putkowski and J. Sykes, *Shot at Dawn,* London, Wharnecliffe, 1989.

*J.J. Pullen, *The Twentieth Maine,* London, Eyre and Spottiswoode, 1959.

G.B. Regan, *Someone Had Blundered,* London, Batsford, 1987.

G.B. Regan, *Saladin and the Fall of Jerusalem,* London, Croom Helm, 1987.

*F. Richardson, *Fighting Spirit,* London, Leo Cooper, 1978.

W. Richardson, *The Epic of Tarawa,* London, Odhams, n.d.

E. Sanger, *Englishmen At War,* London, Alan Sutton, 1993.

N. Schwarzkopf, *It Doesn't Take a Hero,* New York, Bantam, 1993.

S. Snelling, *Gallipoli: VCs of the First World War,* London, Alan Sutton, 1995.

K. Tomasson and F. Buist, *Battles of the '45,* London, Batsford, 1978.

J. Toland, *No Man's Land,* London, Methuen, 1980.

*C. Whiting, *The Battle of Hürtgen Forest,* London, Leo Cooper, 1989.

*C. Whiting, *First Blood,* London, Leo Cooper, 1984.

*B.I. Wiley, *The Life of Billy Yank,* Louisiana University Press, 1992.

*B.I. Wiley, *The Life of Johnny Reb,* Louisiana University Press, 1992.

D. Winter, *Haig's Command,* London, Viking, 1991.

L. Wolff, *In Flanders' Fields,* London, Penguin, 1979.

INDEX

Aggression, 53–54, 68, 70
Alamo. *See* War of Texan Independence
Alcohol use, 65–66, 169
  French, 212–13, 215
American Civil War, 8, 12, 106
  battle of Bull Run, 152
  battle of Chancellorsville, 153
  battle of Fredericksburg, 152
  battle of Malvern Hill, 140
  battle of Missionary Ridge, 143, 151
    Army of the Cumberland, 143–46, 151–52
  battle of Murfreesboro, 153
  battle of Perryville, 152
  battle of Seven Pines, 140
  battle of Shiloh, 138, 152
  and cowardice, 140–41
  seige of Vicksburg, 141
  see also Battle of Gettysburg
American Soldier, (Stouffer), 70
Americans, 108, 150
  softness of, 36, 235, 239, 242, 256
Amicide, 257
Anatomy of Courage (Moran), 4, 5, 15
Anual, fortress of, 211
Appomattox, surrender at, 151
Argonne, the, 24
Artillery, 11, 180, 185, 204
  for demoralization, 186, 198
  on the Somme, 161, 162, 164–66, 169
Australians, 196, 199–200, 246
  in Malaya, 226–32
  in Singapore, 222, 223
Badajoz, seige of, 98, 99
Battle of Acre, 75–76
Battle of Agincourt, 5, 78, 95
Battle of Albuera, 98–105, 124
  British fourth Division, 101–02
  British second Division, 100
  Fusilier Brigade, 102
  Middlesex Regiment, 57, 101
  second Hussars, 100–101
  second Vistula Lancers, 100–101
Battle conditions, 11, 49, 193–94
  Civil War, 149–50, 153
  Crimean War, 116–18, 125–26
Battle fatigue. *See* Combat exhaustion
Battle of Fontenoy, 55, 90, 158
  first Guards regiment, 87–88
Battle of Gettysburg, 127–37
  Alabamians, 129–37
  Cemetary Hill, 127–28, 144, 152
  Little Round Top, 46, 47, 139
  twentieth Maine Regiment, 46, 128–37
Battle of Minden, 55, 90–97, 152
  British victory at, 87, 124
  moral advantage, 158
Battle of Nicopolis, 76–78
Battle of Prestonpans, 88–89, 90
Battle of Stamford Bridge, 66
Battle of Waterloo, 11, 66
  British troops, 98, 127
  Hougoumont, Chateau, 25, 26
Battle of Zorndorf, 6
Beauregard, P.G.T., 141
Belgium, 9, 32, 59, 160

Beresford, General, 99–101, 103
Betts, Edward, Brigidier General, 38–39
Blitzkrieg, 175, 180, 217, 235
Boer War, 158, 170, 190
Bowie, James, Colonel, 108–13
Bradley, Omar, General, 61, 261
Bragg, Braxton, General, 143–44, 147, 148, 150
Bridges, Tom, Major, 48–49
British, 3, 64, 118
  chauvenism, 104, 118–19
  ethnocentrism, 105, 225
  German opinion of, 90
  Grand Fleet, 41
  Johnson, Samuel, on, 104–05
  redcoats, 86–91, 121
  Tommies, 6, 162, 193, 216
  Wolfe, General, on, 89–90
  see also Mercenaries
British Expeditionary Force (BEF), 41, 54, 158, 214, 220
  at St. Quentin, 46–49
Braüchmuller, artillery, 198
Brunswick, duke of, 91–93, 97
Buchnall, General, 22–23, 24
Burma, 49

Campbell, Sir Colin, 46, 47, 120
Cadorna, Italian General, 174
Canadians, 31–32, 61, 89
Capital punishment, 10, 27, 39
  for desertion, 11, 34, 153
  see also Execution
Cardigan, Lord, 47, 65
Central Powers, the, 159, 173
Chamberlain, Colonel, 46, 47
  at Gettysburg, 129, 131–37, 139
Chantilly, allies meet at, 160
"Charge of the Light Brigade," 47, 92, 152, 164, 253
Chattanooga, 143–44, 148
Chemistry, body, 3, 13, 14
Chickamauga, 143–44, 145
Churchill, Winston, 179, 242
  and Singapore, 223–24, 229
Cole, Sir Lowry, 101–02
Collins, Commander "Lightning Joe," 253
Combat breakdown, 2, 3
Combat exhaustion, 66, 181–84
  automatic behavior, 55–56
  battle of the Bulge, 34
  Civil War, 153
  as cowardice, 261
  factors in, 6–7, 11–12, 14
  Gladden, Norman, on, 184
  Holmes, Richard, on, 6
  Hürtgen Forest, 34–36, 254–59
  inevitability of, 6, 11, 68
Commissary, 117
  deficiencies of, 115–18, 150
Compact History of the Civil War (Dupuy), 145
Conscientious objection, 35
  versus duty, 25, 69–70
Conscription, 2, 53–54, 58
  for Civil War, 148–49
Cope, Sir John, 88–89, 90
Cota, Norman "Dutch," 36–39
  Hürtgen Forest, 256, 258–60
Courage, 3–4, 45, 73, 90
  as "a moral quality," 15
  American, 19, 244
  Aquinas, St. Thomas, on, 75
  British, 40, 156, 170
  German, 21–24, 200
  at Gettysburg, 134, 135, 139
  group, 16, 57, 167
  Henry of Ghent on, 75
  individual, 16, 26–27, 78, 122
  Moran's four degrees of, 4, 5
  as pride in race, 105
  through perfectionism, 19, 21
  as will power, 15
Courts-martial, 11, 91, 190
  World War I, 3, 43, 190
  World War II, 33–34, 35, 37, 39
Cowardice, 4, 15, 72–73
  and the breaking point, 11
  in British officer, 40, 43–44, 91
  Civil War, 138–42
  concealment of, 139
  Fuller, Thomas, on, 73
  Ramseur, Stephen, on, 142
  see also Deserters

Crimean War, 60, 115–26, 164
battle on the Alma, 114, 119, 121
battle of Balaclava, 65, 116, 118, 120–21, 127
ninety-third Highlanders, 46, 47, 121
battle of Inkermann, 118, 121–22, 145
Coldstream Guards, 122
fifty-fifth Regiment, 118
fifty-seventh Regiment, 124
Grenadier Guards, 123, 124
Iakoutsk Regiment, 124
twentieth Regiment, 124
British role, 7, 58, 67, 114
London during, 115
Crockett, David, Colonel, 109–10
Cromwell, Oliver, "Ironsides," 8
Crusaders, 76–77, 78, 246
and Antioch, 79, 80, 81
Godfrey of Bouillon, 80–85
Knights Templar, Galilee, 78
Custer's Last Stand, 106, 179

D Day, 30, 244
de Choiseul, foreign minister, 97
de Contades, Marquis Louis, 96
defeat at Minden, 92, 97
de Cos, General, 107, 108, 112
de Poyanne, Marquis, 96
de Ridefort, Gérard, 78
"Desert Rats," 21
Deserters, 3, 198–99
American, 28, 237–38, 240–42
Hürtgen Forest, 258–60
Anual, Morocco, 209–11
Civil War, 138, 140, 149–53
Hill, D.H., General, on, 140
identification of, 4
Italian, at Caporetto, 174–75
Missionary Ridge, 143
on the Somme, 44, 189
see also Cowardice
"The Diehards," 57, 101, 124
Discipline, 13, 186
British, 2–3, 15, 90–91, 221
battle of Minden, 95
Crimean War, 121
broken at Singapore, 222
for democracies, 2, 257
enforcement of, 38–39, 44
French army, 76, 214–15
lack of, 140–41, 152, 230–31
as self-control, 2, 3, 13
Texan settlers, 108–09
see also Military training
Disease, 7, 49, 116–17, 257
Disenchantment (Montague), 59
Duty, 64, 75, 120, 186, 256
to country, 25, 57–58, 60, 69
at Balaclava, 120
German heroism and, 19
to "King and Country," 9
soldier's, 36, 37, 43, 175
to unit or self, 60, 61
Dyett, Edwin, 15, 27, 41–44

Early, Jubal, General, 142
on skulkers, 140, 141
Eisenhower, Dwight D., 235
and Slovik, 37, 38, 39, 40
soldier's rights, 236, 242
Erskine, Major General, 21, 24
Esprit de corps, 182
as adaptation, 67
and courage, 27, 105
essential to morale, 260
French, Sergeant, on, 183
lack of, in young, 192, 197
loss of, by dilution, 90–91
and "regimental system," 9, 51, 56–57, 121
Execution, 3, 153, 261
American soldier, 24–28, 39–40
British naval officer, 40–44
see also Capital punishment

Falklands "Malvinas" War, 57
Fanaticism, 8, 78–83, 246
Fannin, Colonel, 110, 111
Fear in battle, 4, 5, 27
at Albuera, 99
Civil War soldiers, 12–13
control of, 1, 3, 60
see also Military training
Little Round Top, 130

Fear in battle (*continued*)
pervasiveness of, 5, 27
Tarawa atoll, 251
"Fight or flight," 3, 75, 120
syndrome, 13–14, 45
Fighting spirit, 46, 79, 185
American, 239–41, 252
Civil War, 127, 143
British, 104, 196, 200–201
in the Crimea, 114, 118–19
cultivation of, 53, 66
Montgomery, Field Marshal, on, 45
rediscovery of, 47, 199
and revenge, 93, 94, 113
Rudder, James, Lieutenant Colonel, on, 36
as sense of duty, 176
stimulation of, 8, 179
Xenophon on, 45
see also Esprit de corps; Morale
Fighting spirit, breakdown in, 9, 67
American, 237–39, 240
and boredom, 213–14
British, 24, 192–93
Pals Battalions, 157, 158
Italian, 173–77
Russian gunners, 6
Fitzjames, French General, 94–95
Flanders, 162, 185, 187
Foch, French General, 202
France:
as battleground, 160
the "Phony War," 212, 221
Third Republic, 212–13
Fredendall, General, 239–40
Freedom, 106, 110
French, the, 5, 7, 215
and ethnocentrism, 76
Foreign Legion, 52
and "French leave," 215
and mutiny, 3, 180
French, Sir John, 46

Gallipoli, 15, 16, 41, 160, 222
Gamelin, Maurice, General, 219
Geneva Convention, 179
Germans, 3, 8, 21–22, 90
Gerrish, Theodore, Private, 133, 134–35
Goebbels, Joseph, 216
Gough, Sir Hubert, General, 18
the Kaiser battle, 180, 187
Graham, James, Sergeant, 25, 26
Granger, Gordon, General, 146
Grant, Ulysses, General, 152, 153
Missionary Ridge, 143–44, 145, 152
Grimes, General, 142
Grotte, Captain, 32, 33
Guderian, Heinz, General, 216–17
Guibert, Comte de, 56
Gustavus Adolphus, King, 52

Haig, Sir Douglas, Field Marshal, 181, 184–85, 190, 197, 199
first battle of Ypres, 46
and military discipline, 44
and the Somme, 158–59, 160–66, 172
Hankey, Major, 46–47
Hardinge, Henry, Major, 101
Hay, Charles, 87–88
Henry V (Shakespeare), 5
Heroism, 2, 11, 16, 17, 90
German concept of, 18
High Command,
American, 262
French, 215, 216
German, 20, 185
Eastern Front, 180
Italian Front, 174
Highlanders, 55, 87, 90, 191
Balaclava, 46, 120–21
Culloden, 89
the Kaiser Battle, 190–91, 192
Hitler, Adolf, 243
on French army, 212, 217
in Stalingrad, 179
Holmes, Richard, 6, 60–61
Honor, 59–63, 60, 64, 191, 246
Holmes, Richard, on, 60–61
Marshall, S.L.A. on, 61
Houston, Sam, 108–09, 113
Huns, 9, 159, 201, 203
see also Germans

Hussars, 53, 90, 100, 102
  Russian, 120–21
Hutziger, General, 219–20

Indians, American, 55, 86, 90–91
  Sioux and Cheyenne, 106, 179
Inglis, Colonel, 57, 101
Iron Cross, German, 19–20
Italy, 3, 173–75
Iwo Jima, 245

Jacka, Albert, 15–19
  Rule, E.J., Sergeant, on, 16, 18
Jackson, "Stonewall," 152–53, 240
Jacobite Rising, 88
Japanese, 180, 223, 227, 245–46
  Allen, Louis, on, 225
  at Malaya, 222, 224–25
  at Tarawa atoll, 245–52
Jerusalem, 75, 80–85
Joffre, Joseph, Marshal, 160
Johnny Reb (Wiley), 139, 141, 148, 151
Johnson, Albert Sidney, General, 138, 140

Kasserine Pass, defense of, 234, 239–42
Keegan, John, historian:
  on bombardment of defenses, 165
  on Kitchener's volunteers, 154, 155, 157
Keichi Shibasake, Rear Admiral, 247, 249
Kiggell, Sir Lancelot, Haig's Chief of Staff, 163
Kingsley, General, 97
Kingsman, Howard, Admiral, 247
Kitchener, Field Marshal, 59, 163
  see also Pals Battalions
Korean War, casualties, 170

Lafontaine, General, 217–18
Latham, Matthew, Lieutenant, 102
Law, General, 129, 131
Leadership, 24, 49, 54, 87, 210
  confidence in, 9
  incompetent, 87, 174
  inspirational, 47
  qualities of, 19
Lee, Robert E., General, 15, 127, 139, 148, 149
Lincoln, Abraham, President, 11, 147
Lloyd George, David, 184
Longstreet, General, 128, 143
Lookout Mountain, 138, 143–44
Loos, 160, 204
Looting, 13, 64–65, 215, 230
  French knights, 76
  Germans in France, 193–96
Loyalty, 52, 64, 68–69
  to Kaiser, oath of, 51
  regimental, 52

Macbean, Captain, 96
Mainwaring, Lieutenant Colonel, Fusiliers, 49
Malaya, 222–27
  Barber, Noel on, 226
Marne, American troops at, 204
Marshall, S.L.A., 154
McQuillan, General, 237
Meade, George, General, 127–28, 129, 139
Melcher, H.S., Lieutenant, 135
Mercenaries, 52, 62–63, 64
  Lawrence, Sir John, on, 62
Mexico,
  de Iturbide, Agustin, Emperor, 107
  see also War of Texan Independence
Middlebrook, Martin, 191–92
Middleton, Drew, American correspondent, 220
Military Cross, British, 18
Military history, 5, 6, 9, 15
  American Official History, 241
  Australian, Bean, CEW, 18, 199
  British:
    Fortescue, Sir John, 5–6, 88
    Official History, 182
  German, Dinter, Elmar, 13, 14
  regimental, 56–57, 139
  War Diary of Reserve Army, 18
Military training, 10, 162, 214
  depersonalization, 51, 52, 55
  drilling, 5, 53, 55–56, 87
  lack of, 141, 226

Military training (*continued*)
for Pals Battalions, 155–56
in peacetime, 54
purpose of, 51, 53, 55, 56
to counteract fear, 4–5, 21
Smyth, Sir John, on, 10
Military virtues, 51, 78, 256
leadership qualities, 19
natural authority, 16
Monongahela River, 87, 90–91
Montcalm, 89
Montgomery, Bernard, General, 21, 24, 45, 236, 261
Morale, 3, 49, 50, 199
maintenance of, 7, 167
Montgomery, on, 45
Patton, General, on, 36
and positive thinking, 67
Richardson, F.N., on, 7
see also Fighting spirit
Morale, breakdown of, 9
American
Civil War, 136, 142, 144, 147–50
Hürtgen Forest, 254–63
Tarawa atoll, 250
Australian, 222
British, 48, 183, 222, 224
the Kaiser Battle, 187
Normandy campaign, 21
French, 212–17
German, 3, 181
Italian, 3, 174
Spanish, 209–10
Morale, determinants of, 7–8, 9, 13, 114
confidence in leadership, 9
expectations, 9
good food, 7, 13, 65, 149–50
good medical service, 7
good training, 9, 53
maintaining general fitness, 13
relief from stress, 8, 68, 184
rest and sleep, 8, 13, 68
Richardson, F.N., on, 7
see also Commissary
Morale, high, 7, 63, 72, 73
American, 109, 110–11, 130, 131
British, Crimean War, 114, 119
crusaders, 78, 79, 83
Santa Anna's army, 110
Morocco, 209
el Krim, Abd, 210–11
see also North Africa
Morrill, Captain, 129, 135–36
Muslims, 78–82, 85, 246
see also Saracens

Napoleon, 12, 45, 98
and *Grande Armée*, 7
Neill, Colonel, 108, 109, 110
Nelson, William, General, 152
Neuve Chapelle, 160
New Guinea, 49
New Zealand, 223, 246
North Africa, 49, 242
American first Infantry Division, 234–41
Sidi Faid Pass, 236–39, 241

Oates, Colonel, 129–30, 131–37
"Old Contemptibles," 54, 197
"Operation Torch," 235
Operation *Waffentreue*, 174
Ottoman Empire, 77

Pals Battalions, 154–58, 163, 167–68, 170, 171
Morgan, George, Private, on, 157
Panzer troops, 23, 30, 217, 220
Passchendaele campaign, 49, 189
and combat exhaustion, 181, 183, 187, 253
Patriotism, 19, 57–60, 126, 150
British, 58–59, 154–55, 159
and conscription, 148
German, 187
Patton, American General, 36, 261
in North Africa, 242
Pearl Harbor, 224
Pershing, General, 203
Personnel selection, 2, 19, 27, 72, 73–74, 156, 263
Petain, Allied commander, 197
Plundering. *See* Looting
Poland, 215–16, 221, 238, 261

Pope, Union army, 148
Pope Urban II, 82, 83
Portuguese, 197–98
Pour la Mérite, 178
Pozières, 17–18, 160, 167
Primozic, Hugo, 19–21
Propaganda, 21, 57, 139, 216
  British, 8–9, 159
Prussia, 6, 55, 72
  Frederick the Great, 54, 86, 87
Pullen, John, 130, 131–32, 133
Pyle, Ernie, American war
    correspondent, 242

Raglan, Lord, British army, 46
Rawlinson, Sir Henry, commander,
    161–66, 171
  British Fourth Army, 158, 160, 163,
    165
Recruitment, 2, 29, 104, 148
  see also Conscription
Regiment, 9, 91, 121
Religion, 8, 79, 106–07
  see also Fanaticism
Research on combat, 6, 67
Rommel, Erwin, General, 24, 242
  Afrika Korps, 235, 236–39
  in Italy, 173–78
Roosevelt, Franklin D., 239
Rosecrans, Union General, 152
  Union Army of the Cumberland,
    143, 145
Russell, Sir Charles, 122
Russians, 3, 6, 65
  in Crimean War, 119, 120–23

Sabine Cross Roads, 153
Sackville, Lord George, 91
  cavelry at Minden, 95, 96, 97
Saladin and the Fall of Jerusalem
    (Regan), 79
Santa Anna, 107, 109–13
Saracens, 76, 78
  see also Muslims
Schlieffen Plan, 47–48
Sebastapol, 114
Sedan, 24, 217–20
  Ruby, General, at, 218
Self-inflicted wounds, 210, 262
  Civil War, 149
  on the Somme, 168
Seven Years' War, 6
Sexual deprivation, 70
Shaw, John, Corporal, 66
"Shell shock," 12, 183
  see also Combat exhaustion
Shenandoah Valley, 141
Sherman, Union army, 143, 144
Shoup, Colonel, 248, 251
Sickles, General, III Corps,128
Silvestre, General, 210–11
Singapore, 179, 222–33, 247, 256
  British Royal Navy, 222, 232
Skulkers, 140, 142, 153
Slovik, Eddie, 15, 27–40, 262
  execution of, 27, 39–40, 261
  at Hürtgen Forest, 260
  as misfit soldier, 4, 10, 264
Smith, Julian, Major General, 247,
    251
Soldiers,
  civilian, 36, 67, 152, 242, 256
    and cultural clash, 68–69, 72–73
  eighteenth century, 5, 52, 86
  and isolation, 12, 124, 209, 254
  limited days under fire, 5
  nineteenth century, 120, 179
  pay for, 167, 209, 213
  and privacy, lack of, 71
  Wavell, General, on, 9, 232
Somner, Colonel, 38
Soult, Marshal, 98, 100–101, 102
Stewart, General William, 100
Storm of Steel (Junger), 189
Stress, 45, 72, 152
  and body chemistry, 3, 13, 14
  imposed by cultural clash, 69
  involuntary reaction to, 13
  morale damaged by, 8, 67
  rationalization of, 68
  and reasoned judgment, 14
  and sexual deprivation, 70
Stuart, "JEB," 148
Sumner, Henry J., Lieutenant
    Colonel, 35
Sykes, General, V Corps, 128

Tanks,
American, 237, 247, 251, 258
British, 22–24, 241
German, 22–24, 237, 239, 240
Japanese, light, 247
Stalin Tank Corps, 20
Tarawa atoll, 245–52
Technology, 12, 225, 255
Tennessee volunteers, 109–10
Texas, San Antonio, 108, 109
settlement of, 106–07
see also War of Texan Independence
"the thin red line," 120
Thirty Years' War, 52
battle of Marston Moor, 53
Thomas, George, "Rock of Chickamauga," 143–44, 145, 146
To Lose a Battle (Horne), 217
Travis, William B., Lieutenant, 109–13
Tunisian campaign, 234, 239–42
first division, American, 239
second Corps, 236–39
tenth Panzer Division, 237
Turks, 16, 76–77, 120, 183

Victoria Cross, 16, 18, 191
Vikings, 65–66, 246
Vincent, Colonel, 128–29, 131
von Below, General, 174, 175
von Kluck, General, 47–48, 54, 158, 191
von Ludendorff, Field Marshal, 199
on British discipline, 2–3
offensive, 50, 179, 200, 235
against British Fifth Army, 180–81, 185
in the Dolomites, 174, 175
von Spoercken, General, 92, 96, 97

Waldegrave, General, 92–93, 95, 96
War of American Independence, 86
Braddock's defeat, 87, 90–91
for a "cause," 106
War of 1812, 108
War in the Pacific, 222–33
British Royal Navy, 222, 224, 232
forty-eighth Battalion, Australian, 222–32
second Marine Division, at Tarawa, 245–52
War of Texan Independence, 106–13
Alamo, seige of, 109–11
battle of San Jacinto, 113
for a "cause," 106, 110–11, 113
Gonzales (settlement), 107–08
Warfare,
eighteenth century, 5, 86, 92, 207
"kill or be killed," 65, 68, 72
nineteenth century, 7, 209
twentieth century, 6, 12
Warren, Gouverneur K., General, 128
Wehrmacht, 19, 61, 235
in France, 30, 34, 39
Wellington, duke of, "Iron Duke," 25, 103, 118, 119
on British soldiers, 89, 98, 104
Peninsular veterans, 54, 105
at Waterloo, 128
Western Way of War (Hanson), 4
Whiting, General, 140
Wilhelm II, Kaiser, 185
Wittman, Michael, Captain, 21–24
World War I, 6, 14
Australian Imperial Force (AIF), 15–16, 17–18, 230
battle of Belleau Wood, 203–08
fourth Marines Brigade, 204–08
second Division, American, 204
battle of Caporetto, 3, 173–78, 183
German Fourteenth Army, 174
Italian Second Army, 174–75
Wurttemburg Mountain Battalion, 175–78
battle of the Frontiers, 204
battle of Le Cateau, 47, 197
battle of Messines, 18
battle of Mons, 47, 54, 119, 158, 197
battle of Polygon Wood, 18
battle of the Somme, 17, 160–72, 204
Australian forty-eighth Battalion, 17–18
British Fourth Army, 158

casualties, 40–41, 157–58, 170, 171, 172, 196
fourteenth Battalion, 16–18
German Imperial Army, 170
Hunter-Weston's Eighth Corps, 172
Indian mercenaries, 63–64
underground defenses, 160, 165
see also Kitchener, Field Marshal
battle for Verdun, 160, 171, 204, 215, 217
battle of Ypres, first, 54, 158, 191
First Army, German, 47–48
Third Army, German, 46
Worchestershire Regiment, 46–47
the Kaiser Battle, 185–93, 198
British Fifth Army, 180–81, 182, 185, 187, 192
British rout at Peronne, 189
Gloucestershires, 191, 192
March Offensive, 65, 197
see also Passchendaele campaign
World War II, 175, 208
D-day, 30, 244
battle of the Bulge, 34, 38, 264
battle of Hürtgen Forest, 32, 33, 36, 253, 257–64
ninth Division, American, 254–55
twenty-eighth Infantry Division, 30–34, 36, 255–60
battle of Rshev, 20
battle of Villers-Bocage, 21–24
fifty-first Highlanders, 21
first Airborne Division, 21
seventh Armored Division, 21–24
SS Heavy Tank Battalion, 501st, 21
XXX Corps, 22–23, 24
Maginot line at Sedan, 24, 212, 214, 216, 221
fifty-fifth Division, French, 213, 217–19
Normandy campaign, 21–22, 24, 244, 253
first Airborne Division, 21
Omaha Beach, 30, 256
seventh Armored Division, 21, 22, 23, 24
see also Tunisian campaign; War in the Pacific

York, Alvin, Corporal, 15, 25